7/1/14
Flaming gorge
Info Center

The Geologic Story of
the Uinta Mountains

The Geologic Story of the Uinta Mountains

Second Edition

Wallace R. Hansen

FALCONGUIDE®

GUILFORD, CONNECTICUT
HELENA, MONTANA
AN IMPRINT OF THE GLOBE PEQUOT PRESS

Photos courtesy of U.S. Geological Survey and USDA Forest Service, unless otherwise credited
Illustrations by John R. Stacy and Clint McKnight

Library of Congress Cataloging-in-Publication Data

Hansen, Wallace R., 1920–
    The geologic story of the Uinta Mountains / Wallace R. Hansen.— 2nd ed.
        p. cm.
    Includes bibliographical references and index.
    ISBN 0-7627-3810-3
        1. Geology—Uinta Mountains (Utah and Wyo.) I. Title.

QE79.5.H35 2005
557.92'14—dc22   2005040069

Manufactured in the United States of America
Second Edition/First Printing

Intermountain
NATURAL
HISTORY
Association

This book has been published in cooperation with the Intermountain
Natural History Association. INHA is a nonprofit organization created
to aid the interpretive, educational, and scientific activities of the
National Park Service at Dinosaur and Fossil Butte National Monu-
ments, the USDA Forest Service at the Ashley, Wasatch-Cache, and
Uinta National Forests, and the Bureau of Land Management at the
John Jarvie Ranch Historic Site.

Contents

Figures

Preface

My earliest recollections of the Uinta Mountains reach back almost to the threshold of memory. As a schoolboy I made many hikes to the high ridgelines of the Wasatch Range, east of Salt Lake City, and on such occasions my father often pointed out the snowy peaks of the High Uintas shimmering far off to the east. Mount Timpanogos, near Provo, was a favorite climb in those days, and from its lofty summit all of northern Utah seemed to be spread out at one's feet—all except the Uinta Mountains, which loomed even higher on the horizon. My first actual visit to the Uintas was with an uncle who was an ardent fisherman and the owner of a 1928 Essex coupe. Mirror Lake, 80 miles from Salt Lake City, was at the end of the road and was a fair test, even for an Essex. A day's limit at Mirror Lake was 30 trout, and catching that many wasn't much of a chore. From time to time my uncle and I tried out other lakes in that vicinity, dozens of which were visible from the summit of Bald Mountain. Scout Lake is one, north of Bald Mountain, at Camp Steiner. At an altitude of 10,300 feet above sea level, Steiner is the highest Boy Scout camp in the United States. It was the ultimate objective of every Boy Scout in the Salt Lake Council. No doubt, it still is.

Later, I made the trip to the Uintas many times on my own, in my 1932 Chevrolet roadster, and extended my horizons to the Upper Weber, Haydens Fork of the Bear, the Duchesne, Rock Creek, and Yellowstone Creek. The hike into the lakes was as much fun as the fishing; the cirque walls, the polished rock surfaces, and the erratic boulders all were rather obvious manifestations of glacial erosion, even to an untrained eye.

In 1940 I visited the Uinta Mountains as a student. Under the direction of R. E. Marsell and the late Hyrum Schneider, the University of Utah had set up a geologic field camp on the Duchesne River. Working from there, I gained my first real appreciation of the country's impressive geology.

After a dozen years I returned to the Uintas in the early 1950s, this time in the employ of the U.S. Geological Survey, to map a block of quadrangles along the Green River between Flaming Gorge and Browns Park. I gained a further appreciation of Uinta Mountain geology and made lasting friendships with many local people.

Over the years I have become acquainted with most of the Uinta Mountains, to some degree, and the following report is drawn mostly from my own notes and memory; but the observations and conclusions of others have been freely utilized. Consciously or unconsciously, I've borrowed from the ideas and published works of J. D. Sears, W. H. Bradley, D. M. Kinney, and other colleagues of the Geological Survey, and from my friends Ernest and Billie Untermann, codirectors of the Utah Field House of Natural History at Vernal and widely recognized authorities on the geology of the Dinosaur National Monument area. Other authors have been cited where appropriate, but no attempt has been made to exhaust the published literature.

Many statements on the following pages perhaps seem assertive. In a more formal report these statements would be carefully documented. But the geologic literature on the Uinta Mountains reaches back nearly 140 years and is voluminous. Fuller documentation would be distractive in a report of this sort. The few reports cited here contain extensive bibliographies of their own, and the reader is referred to them for more background literature.

The intended scope of the report is fairly broad, but the treatment is not comprehensive. Indeed, in a report this size, only the barest outline can be given on the physical traits of an area as geologically encompassing as the Uinta Mountains.

The Uinta Mountains

*A geologic review commemorating John Wesley Powell's
historic explorations among the lofty peaks and deep
canyons of the Uinta Mountains*

Turreted Red Castle, altitude 12,825 feet, stands above a glacial lake on the north flank of the Uinta Mountains.

"*Away to the south, the Uinta Mountains stretch in a long line; high peaks thrust into the sky, and snow-fields glittering like lakes of molten silver; and pine forests in a somber green; and rosy clouds playing around the borders of huge, black masses; and heights and clouds, and mountains and snow-fields, and forests and rocklands, are blended into one view.*"

Major John Wesley Powell's diary, May 24, 1869

"Above all it is a rocky region; rocks are strewn along the valleys, over the plains and plateaus; the cañon walls are of naked rock; long escarpments or cliffs of rock stand athwart the country, and everywhere are mountains of rock. It is the Rocky Mountain region."

Major John Wesley Powell, 1876
Report on the Geology of the Eastern Portion of the Uinta Mountains

Introduction

The opening of the West after the Civil War greatly stimulated early geologic exploration west of the 100th Meridian. One of the areas first studied, the Uinta Mountains region, gained wide attention as a result of the explorations of three Territorial Surveys, one headed by John Wesley Powell, one by Clarence King, and one by Ferdinand V. Hayden. Completion of the Union Pacific Railroad across southern Wyoming in 1869 materially assisted geologic exploration, and the railheads at Green River and Rock Springs greatly simplified the outfitting of expeditions into the mountains.

The overlap of the Powell, King, and Hayden surveys in the Uinta Mountains led to efforts that were less concerted than competitive and not without acrimony. Many parts of the area were seen by all three parties at almost the same time. Duplication was inevitable, of course, but all three surveys contributed vast quantities of new knowledge to the storehouse of geology, and many now-basic concepts arose from their observations.

Canyon of Lodore

Powell's area of interest extended mainly southward from the Uinta Mountains to the Grand Canyon, including the boundless plateaus and canyons of southern Utah and northern Arizona. King's survey extended eastward from the High Sierra in California to Cheyenne, Wyoming, and encompassed a swath of country more than 100 miles wide. Hayden's explorations covered an immense region of mountains and basins from Yellowstone Park in Wyoming southeast throughout most of Colorado.

Powell first entered the Uinta Mountains in the fall of 1868, having traveled north around the east end of the range from the White River country to Green River, Wyoming, then south over a circuitous route to Flaming Gorge and Browns Park, and finally back to the White River, where he spent the winter. In 1869, after reexamining much of the area visited the previous season, Powell embarked on his famous "first boat trip" down the Green and Colorado Rivers. This trip was more exploratory than scientific; his second, more scientific trip was made 2 years later. Powell revisited the Uinta Mountains in 1874 and 1875 to complete the studies begun 6 years earlier. His classic "Report on the Geology of the Eastern Portion of the Uinta Mountains and a Region of Country Adjacent Thereto" was published in 1876.

King's survey—officially "The United States Geological Exploration of the Fortieth Parallel"—is better known simply as the "40th Parallel Survey." King began working eastward from California in 1867. The Uinta Mountains region, however, was mapped by S. F. Emmons, under the supervision of King, in the summers of 1869 and 1871. Emmons' work was monumental, and although he emphasized in his letter of transmittal to King the exploratory nature of the work—as the formal title of the report indicates—his maps, descriptions, and conclusions reflect a comprehensive understanding of the country and its rocks. The 40th Parallel report contains the best, most complete early descriptions of the Uinta Mountains. It, indeed, is a treasure chest of information and a landmark contribution to the emerging science of geology.

Hayden visited the Uinta Mountains in 1870, descending the valley of Henrys Fork to Flaming Gorge in the fall after having earlier examined the higher part of the range to the west. Most of Hayden's observations were cursory, and he repeatedly expressed regret at having insufficient time for more detailed studies. In reference to the area between Clay Basin and Browns Park, he remarked (Hayden, 1871, p. 67) somewhat dryly that "the geology of this portion of the Uinta range is very complicated and interesting. To have solved the problem to my entire satisfaction would have required a week or two." Eighty-odd years later I spent several months there looking at the same rocks.

Powell was perhaps more creative—more intuitive—than either King or Hayden, and his breadth of interest in the fields of geology, physiography, ethnology, and conservation was enormous. King's work, however, was distinctly more disciplined and reflected both his early professional training at Yale, perhaps, and the professional acumen of his subordinates. Powell was largely self-taught; Hayden was formally trained as a surgeon. Powell's reports are almost lyrical, but King's are more erudite, better organized, and more complete as to details than either Powell's or Hayden's. Hayden's reports are chaotic, possibly because he demanded and attained early publication of his findings. His practice was to publish annual reports and, to his lasting credit, most of his results appeared in print within a year of his fieldwork. So far as the Uinta Mountains are concerned, Hayden's thinking was more modern and more sophisticated than that of either of his colleagues.

One example serves to illustrate: The Uinta Mountain Group (Precambrian) rests on the Red Creek Quartzite (older Precambrian) with an angular unconformity of considerable local relief. A local relief of several tens of feet is evident in many places, and at several places it is more than 50 feet. At one place, the relief may be several hundred feet in a horizontal distance of about a mile. The early geologists who visited the area, especially Powell, made much of the supposed great magnitude of the unconformity. Powell deduced 8,000 feet or more of exposed relief, observing correctly the sharp local relief but apparently mistaking a fault to be an unconformable contact. Extrapolating further, Powell visualized that a lofty headland 20,000 feet high had been buried beneath the sediments of an ancient Uinta sea. King concurred. But Hayden's view (1871, p. 66) was much more modern: "I am inclined to believe that the immense thickness of quartz [Red Creek] was thrust up beneath the red quartzites [Uinta Mountain Group] carrying the latter so high up that they have been swept away by erosion." The Red Creek Quartzite, we now know, was indeed uplifted relative to the Uinta Mountain Group by faulting. Although Hayden spent but a month exploring virtually the whole north flank of the Uinta Mountains, from the Bear River on the west to Browns Park on the east, he had an unmistakable grasp of the geology.

Other geologists of note visited the Uinta Mountains in the closing decades of the 19th century, about the time of the Territorial Surveys or shortly thereafter. In 1870 O. C. Marsh, a leading paleontologist of his day, began a study of the Tertiary vertebrates of the Green River Basin, particularly the rich faunas of the Bridger Formation. Marsh's study was made in open competition with E. D. Cope of the Hayden Survey. Marsh, of Yale University, was assured early publication in the *American Journal of Science* published at Yale. Cope therefore resorted to telegraphing fossil descriptions to his publisher to protect his priority of discovery (Merrill, 1906, p. 598).

Marsh visited the Uinta Mountains briefly in the fall of 1870 and published a short description of his travels the following spring (Marsh, 1871, p. 191). Marsh Peak, looming above Vernal, bears his name. Heading south from Fort Bridger, Wyoming, Marsh followed Henrys Fork to its mouth. He turned east at Flaming Gorge to Browns Park, then went south to the Uinta Basin, staying well west of Lodore Canyon, via what is now Pot Creek and Diamond Mountain Plateau. Returning to Fort Bridger, he recrossed the mountains by way of the Uinta River and Sheep Creek Gap, probably skirting North and South Burro Peaks. From some point in his descent of the north flank of the Uinta Mountains, Marsh reported that "while descending the northern slope of the mountains toward the great Tertiary basin of the Green River, which lay in the distance, 2,000 feet below us, we passed over a high ridge, from the summit of which appeared one of the most striking and instructive views, of geological structure to be seen in any country. Sweeping in gentle curves around the base of the mountains, from near where we stood, many miles to the northward, was a descending series of concentric, wavelike ridges, formed of the upturned edges of different colored strata, which dipped successively away from the Uintahs; those nearest to us, 40° or more, those at a distance, seemingly but little—altogether a scene never to be forgotten." Marsh could only have been looking north toward Lucerne Valley, from Windy Ridge above Sheep Creek, with Jessen Butte to his left and Flaming Gorge to his right; this view is repeated nowhere else on the north flank of the range. Sheep Creek Gap, which, in Marsh's words (1871, p. 197), was "a narrow and almost impassable side ravine," now contains paved Utah State Highway 44.

Unusually rich vertebrate faunas in the Bridger and Uinta Formations attracted other paleontologists, among them Joseph Leidy, W. B.
Scott, H. F. Osborne, and Earl Douglass. Leidy, whose namesake is Leidy
Peak, south of Manila, collaborated with King. Scott and Osborne represented Princeton University. Later, in 1909, Douglass, of the Carnegie
Institution, discovered the famous dinosaur bone deposits north of
Jensen, Utah, in what is now Dinosaur National Monument. These paleontologists were interested primarily in the fossil-bearing beds of the
basins, particularly those in the badlands where rock exposures are max-
imal, concealment by soil or vegetation is minimal, and the chances of
finding fossils are optimal. If they visited the forested valleys and ridges
of the high mountains, they made little note of it, though after a hot day
of digging in the badlands, they must have glanced wistfully at the snow-
flecked peaks.

C. A. White reviewed the geology and physiography of the Eastern
Uinta Mountains in 1889, using, of course, much from the works of Powell, King, and Hayden. White described the Eastern Uintas in some detail,
particularly the folded rocks at the flanks of the range, calling attention to
the unusual relations of geologic structure to drainage in the remarkable
canyon country of the Green and Yampa Rivers. How, questioned White,
were streams that were flowing in open valleys on both sides of the range
able to establish and maintain courses across the mountains? Like Powell,
with whom White had been associated in the field, White attributed the
relations to "antecedence"—though he did not use the term—to a belief
that the drainage preceded the folding and, through vigorous erosion,
maintained its course as the mountains rose beneath it. In 1877 Emmons
had expressed a contrary view. He believed that the course of drainage
across the Uinta Mountains was superimposed from an earlier, simpler
terrain—later eroded away—onto the present complex one. His view, with
modifications, is favored here. The details of this bold concept are outlined
on a later page. In brief, a stream initiates its course in a particular geologic
setting, such as a sloping alluvial plain. Then, as it erodes downward it
uncovers new, perhaps adverse, rock conditions at depth, but—being now
held in by canyon walls of its own making—it is compelled to remain in
its established course. "The sinuousness and irregularity of the course of
the Green River through the Uinta Range, and its independence of the

present topographic features of the region," said Emmons, "suggest that it must have been determined originally in rocks of an entirely different nature from those, through which it now lies."

C. R. Van Hise visited the Uinta Mountains in 1889 to look at the Precambrian core of the range. Van Hise was a leading authority on the Precambrian rocks of the United States, and he (1892, plate 8) showed insight in assigning the Uinta Mountain Group to the Precambrian at a time when most geologists still considered it to be Paleozoic. Powell apparently had already reached a similar conclusion as early as 1877, according to White (1889, p. 687).

The impact of the Uinta Mountains on early geologic thinking is hinted at by the names that were given to the great peaks. Most mountain ranges, by the names of their summits, honor the memories of statesmen, politicians, or explorers. The Uintas honor geologists. But to be a successful geologist in the third quarter of the 19th century, one perhaps had also to be a statesman and a politician, and certainly an explorer. Powell was all of these, and he excelled at each. Strangely, no Uinta peak bears his name, although he himself named many landmarks—mountains, streams, and canyons—throughout the Uinta Mountains and the Colorado Plateau.* Powell's names are inclined to be imaginative and descriptive, if not downright dramatic—Flaming Gorge, Kingfisher Canyon, Beehive Point, Swallow Canyon, Canyon of Lodore, Echo Park, Dirty Devil River, Music Temple, and Bright Angel Creek, to mention but a few. He also changed the name of Browns Hole to Browns Park.

No less than a dozen major Uinta Mountain summits bear the names of early-day geologists or topographers—Hayden, Agassiz, Wilson, Gilbert, Emmons, and King among them. Kings Peak, at an altitude of 13,528 feet, is the highest point in the state of Utah. Kings Peak, incidentally, is not on the main divide of the Uinta Mountains but rises from a subsidiary ridge a mile to the south.

---

* After this bulletin was written, I submitted the name "Mount Powell" to the Board of Geographic Names, for a previously unnamed 13,159-foot peak at the crest of the Uinta Mountains just east of Red Castle, between Henrys Fork and the East Fork of Smith Fork. This name has since been approved and will appear on maps published by the government in the future. Also, the new 7½-minute quadrangle map on which Mount Powell is shown is designated the "Mount Powell quadrangle."

Geographic Setting

Geographers divide the Rocky Mountain System in the Western United States into three parts: the Northern, Middle, and Southern Rocky Mountains. The Middle Rocky Mountains, of which the Uintas are a part, extend southward from southern Montana, through eastern Idaho and western Wyoming, into central Utah and include such ranges as the Teton, Wind River, Salt River, and Wasatch. The Uintas are bounded on the north (fig. 1) by the Green River Basin, the Rock Springs uplift, and the Washakie and Sand Wash Basins. They are bounded on the south by the Uinta and Piceance Basins, which are separated by the Douglas Creek arch. Thus, the Uinta Basin is the counterpart of the Green River Basin; the Douglas Creek arch is the counterpart of the Rock Springs uplift; and the Piceance Basin is the counterpart of the Washakie and Sand Wash Basins. The Uinta Basin, the Douglas Creek arch, and the Piceance Basin are components of the Colorado Plateaus province; hence, the south margin of the Uinta Mountains also coincides with the north margin of the Colorado Plateaus.

The Uinta Mountains have an overall length of about 150 miles and a mean width of about 35 miles. At their widest, toward the west, they are more than 45 miles across, and at their narrowest, near the center of the range, they are less than 30 miles across. The boundaries are somewhat indefinite, inasmuch as the flanks pass into bordering hogback ridges and broad sloping mesas that merge gradually with the high arid tablelands of the adjacent basins.

The Uinta Mountains have been carved from an immense anticlinal uplift, an upward flexing of the earth's crust which nearly coincides with the outer limits of the range. This great fold is described in some detail on the pages that follow. Down through countless centuries, the fold has been attacked and dissected by the elements, and in its present eroded form, it is largely responsible for the overall character of the range.

For a better understanding of the physical and spatial relations of the Uinta Mountains, topographic maps are an invaluable aid. Good topographic coverage is available at small scales for the entire range, and at modern large scales for most of it. Modern detailed maps by the U.S. Geological Survey are published at a scale of 1:24,000 (2,000 feet per

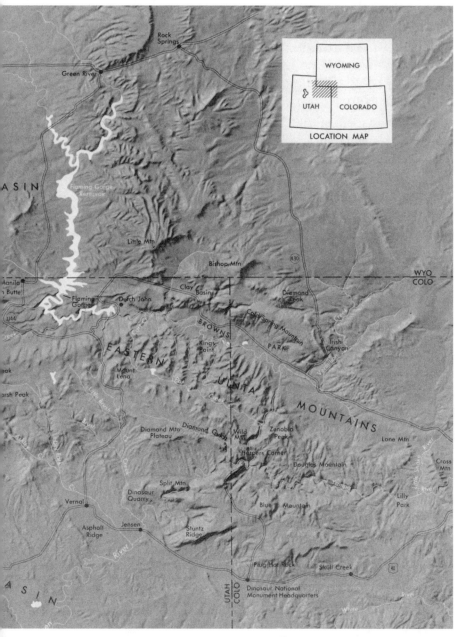

The Uinta Mountains (Fig. 1)

inch). These maps, with a green forest overprint, show roads, trails, buildings, lakes, meadows, and springs, and they identify most of the named landmarks. The old Coalville (1900), Hayden Peak (1901), Gilbert Peak (1905), and Marsh Peak (1906) quadrangle maps, at a scale of 1:125,000 (2 miles per inch), show graphically most of the west half of the range, although most of the altitudes have been revised with modern mapping, and the culture is out of date. These fine old maps are slowly disappearing from stock and soon will be collector's items. Regional relations are very well shown by the 1:250,000 (4 miles per inch) Salt Lake City and Vernal topographic sheets, which together portray the entire range. These maps were prepared by the U.S. Army Map Service and are published in civil edition by the U.S. Geological Survey. Plastic raised-relief editions of these maps, available from the U.S. Army Map Service, depict the physiography exceedingly well in three dimensions and enable the user to see the form of the mountains in a way not otherwise possible. Excellent forest maps of the Wasatch and Ashley National Forests are available from the USDA Forest Service. These maps show roads, trails, peaks, streams, and lakes and include a large part of the Uinta Mountains.

The Landscape and its Attributes

Timberline in the Uinta Mountains is at an altitude of about 11,000 feet. All the high peaks and ridges above that altitude, from Mount Watson on the west to Marsh Peak on the east, are in the west half of the range. The range, in fact, can be divided topographically as well as structually into west and east counterparts, here referred to as the Western and Eastern Uinta Mountains, respectively, or more simply, as the Western and Eastern Uintas. The often-used name "High Uintas" is synonymous with the Western Uinta Mountains. A broad low pass north of Vernal, drained by Cart Creek on the north and by a tributary of Brush Creek on the south, separates the two parts of the range.

Arête. Glacially sharpened Red Castle, viewed from the west, dominates skyline above Smith Fork just north of the Uinta crestline. Massive flat bedding characterizes the Uinta Mountain Group (Precambrian) in this area. (Fig. 2)

Glacial landscape high on the south flank of the Uinta Mountains. Unnamed ice-sculptured peak, altitude 12,385 feet, at head of Rock Creek Basin, looms above morainal ridge in foreground. Photograph by Max D. Crittenden, Jr. (Fig. 3)

Felsenmeer. Jumble of frost-riven blocks tops ridges unreached by Ice Age glaciers. Valley of Yellowstone Creek, viewed toward the northeast. Compare with figures 2 and 3. Photograph by Max D. Crittenden, Jr. (Fig. 4)

The Western Uinta Mountains

Peaks and Ridges

The main divide, or "backbone," of the Western Uinta Mountains is a narrow sinuous ridge more than 60 miles long and rarely as much as a mile across at its base, extending from Hayden Peak on the west to Leidy Peak on the east. Until late Tertiary time (2–15 mya) this ridge probably was the Continental Divide. The base of the ridge stands near or somewhat above timberline, 11,000–12,000 feet above sea level. The crest is 1,000–2,000 feet higher. From the head of Stillwater Fork, east of Mount Agassiz, to Island Lake, east of the Burro Peaks—a distance of about 43 miles—the base of the ridge is nowhere less than 11,500 feet above sea level; most of it is higher than 12,000 feet. North and south from the main divide are many subsidiary spurs and ridges, most of them equally narrow, equally sinuous, and as high or higher than the main divide. The new topographic maps of the Uinta Mountains show 26 individual summits, including some subordinate peaks, that stand more than 13,000 feet above sea level; however, only nine of these are on the main divide.

Particularly near the crest of the range, the summit ridges are mostly steep-sided forms that alpinists would call arêtes. Exposed ledges of bedrock, mantled only here and there by talus, are surmounted by little or no flat space at the ridgeline (figs. 2 and 3). Subordinate ridges between canyons, on the other hand, commonly broaden away from the crest into lofty plateaus 10,000–12,000 feet above sea level. These ridges, and wide places along the crest itself, are covered with loose angular rubble—a vast felsenmeer, or "block field," subjected since Tertiary time to the relentless attack of frost and water (fig. 4). Perhaps most of this rubble is a product of the past severe climate of the Pleistocene Epoch, or Great Ice Age (10,000 years–2 mya). At that time, the valleys below were buried repeatedly under great accumulations of moving ice, which swept away all loose rock and redeposited it downstream. But the exposed ridges, standing high above the ice, were under almost continuous attack by frost.

Frost has a ratcheting effect on rocks. Water percolates into an opening, freezes and expands, widens the opening, thaws, percolates deeper, refreezes, widens the opening further, and so on. Eventually, solid rock is split asunder. The process is effective at high altitudes throughout the world and in the polar regions. It is particularly effective on brittle jointed quartzite, like that in the High Uintas, where moisture has ready access to cracks and bedding planes. Even today the process there is very active, inasmuch as freezing temperatures can occur in any month of the year. The effect of the process is to soften the sharp outlines left by the ice—the arêtes, headwalls, and cirque basins are all being modified by frost, snowmelt, gullying, and wind. If the present process continues long enough, the steep-sided ridgelines will eventually become mere rubble-covered mounds, buried in their own debris.

Drainage Basins

The summit ridges of the Western Uintas form the watersheds of several major drainage basins. Water falling on the north flank drains to the Great Basin via the Weber and Bear Rivers, which empty into the Great Salt Lake, and to the Green River Basin via Blacks Fork of the Green, Smith Fork, Henrys Fork, Beaver Creek, Burnt Fork, Sheep Creek, and Carter Creek. The south flank is drained by the Provo, which flows to Utah Lake in the Great Basin, by the Duchesne, which flows to the Green River in the Uinta Basin, and by such tributaries of the Duchesne as Rock Creek, Lake Fork, Yellowstone Creek, the Uinta River, and the Whiterocks River. Ashley Creek and Brush Creek, near Vernal, flow independently to the Green River in the Uinta Basin.

Sculpture by Moving Ice

All these streams head near the crest of the range in broad amphitheater-like basins, or compound cirques, some of extraordinary size, scoured out by now-extinct Pleistocene glaciers. Lying near timberline, most of these basins are floored by glacially abraded bedrock mantled only in part by soil and vegetation. In aggregate, the basins contain hundreds of lakes and ponds—some in hollows eroded from solid rock, others dammed by glacial moraines (fig. 5). Many lakes are arranged in stair-step fashion, one above another; the seven Chain Lakes at the foot of Mount Emmons

Glacial lake in the Uinta Drainage. Hundreds of such lakes were left in the Uinta Mountains in the wake of the melted Ice Age glaciers. (Fig. 5)

and the Red Castle Lakes of Smith Fork are good examples. In a setting of deep forests and alpine meadows the silent beauty of these lakes enhances the lonely splendor of the mountains. Emmons, in 1877, duly impressed by what he had seen, reported that "the scenery of this elevated region is singularly wild and picturesque, both in form and coloring. In the higher portions of the range, where the forest-growth is extremely scanty, the effect is that of desolate grandeur; but in the lower basinlike valleys, which support a heavy growth of coniferous trees, the view of one of these mountain lakes, with its deep-green water and fringe of meadowland, set in a sombre frame of pine forests, the whole enclosed by high walls of reddish-purple rock, whose horizontal bedding gives almost the appearance of a pile of Cyclopean masonry, forms a picture of rare beauty." A few lakes can be reached by automobile, especially those in the Mirror Lake–Bald Mountain area on Utah State Highway 150. But most of them are accessible only on foot or horseback, just as in Emmons' time.

Extent of the Ice

The great amphitheaters along the crest of the range extend basinward into long steep-sided canyons 2,000–3,000 feet deep. During the Great Ice Age these canyons contained large trunk glaciers that moved outward from their catchments in the amphitheaters toward their termini near the mountain flanks (fig. 8). The form of the amphitheaters and canyons, in fact, is due to the scouring action of the ice on their walls and floors. At one time some of the glaciers extended well beyond the mountains onto the adjacent plains. Several were more than 20 miles long; the longest known was the Blacks Fork glacier on the north slope, which, according to Bradley (1936, plate 34), deposited its outermost moraine on the plains of the Green River Basin 38 miles north of the crest. At an early glacial stage several glaciers on the north slope merged into broad piedmont ice sheets. The longest known glacier on the south slope was the Uinta River glacier, which was about 27 miles long. However, the merged Lake Fork and Yellowstone Creek glacier was more massive. Altogether, the ice at its maximum extent covered at least 1,000 square miles, according to W. W. Atwood (1909), who first studied in detail the effects of the glaciers on the Uinta Mountains. No doubt, the ice was much more extensive at one time than Atwood estimated because he based his estimate on the extent of relatively late Pleistocene glaciers, which were confined largely to the valleys. Atwood was unaware of the far-reaching older glaciations, discovered later by Bradley, or of extensive piedmont deposits more recently mapped by the Geology Department of the University of Utah (Stokes and Madsen, 1961). Perhaps the total extent of the glaciers was closer to 1,500 square miles.

Atwood identified two distinct "epochs" of glaciation and suggested a third. Bradley (1936), although he made no attempt to study the glaciations systematically, identified three stages, including one considerably older than Atwood's "earlier glacial epoch." This, he called the "Little Dry Stage" for the morainal deposits near Little Dry Creek south of Mountainview, Wyoming. Richmond (1965) recognized two additional

Atwood (1909)	Bradley (1936)	Richmond (1965)		Years Ago	Character
		Neoglaciation	Gannett Peak Stade		Small cirque glaciers. Fresh, young moraines just below the cirque headwalls; protalus ramparts, and rock glaciers that support no vegetation.
			Temple Lake Stade	800–900 / 4,000	Cirque glaciers. Fresh young moraines on or near floors of cirques. Rock glaciers. Scanty vegetation.
Later glacial epoch	Smith Fork Stage	Pinedale Glaciation	Late stade	6,500 / 10,000	Valley glaciers. Massive bouldery, fresh-looking moraines, both terminal and lateral. Striations, flutings, and polish on bedrock.
			Middle stade		
			Early stade	25,000 / 32,000	
Earlier glacial epoch	Blacks Fork Stage	Bull Lake Glaciation	Late stade	45,000	Valley glaciers. Massive terminal moraines, weathered and eroded.
			Early stade		
	Little Dry Stage	Pre–Bull Lake glaciations			Valley and piedmont glaciers. Deeply eroded deposits. Former extent not fully known.

Correlation chart showing glaciations of the Uinta Mountains. (Fig. 6)

Ice Age glaciers of the Uinta Mountains. Maximum known and inferred extent of the ice. Piedmont glaciers shown on the north flank of the range existed only in earlier glacial stages. (Fig. 7)

younger "stades" high in the mountains and subdivided Atwood's two "epochs" into five "stades" (fig. 6).

The effects of the glaciers on the Uinta Mountains are still clearly visible today, though many streams have since incised themselves into narrow inner gorges. V-shaped canyon profiles, truncated ridge spurs, hanging tributary canyons, falls and cataracts, and polished rock surfaces all testify to the abrasive action of moving ice (fig. 9).

Although the glaciers swept the upper reaches of many canyons nearly clean of loose rock, they deposited hummocky moraines in most of the downstream reaches. Some of these moraines are well shown on the new topographic maps, particularly those of the Taylor Mountain, Whiterocks Lake, Leidy Peak, Hole-In-The-Rock, and Gilbert Peak Northeast quadrangles. The Taylor Mountain quadrangle map shows massive looped moraines crowded with hundreds of little kettle ponds along the South Fork of Ashley Creek. The Whiterocks Lake and Leidy Peak quadrangle maps show several sequences of downstream moraines on the forks of Sheep and Carter Creeks, as well as younger moraines high in the cirques and many cirque lakes. The Hole-In-The-Rock quadrangle map shows massive moraines of Bull Lake and Pinedale age and broad outwash plains along the West and Middle Forks of Beaver Creek. The Gilbert Peak Northeast quadrangle map shows a complex of large pitted moraines along Henrys Fork and an unusual constriction of the valley where the glacier flowed between resistant ramparts of Mississippian limestone.

Many canyons are choked with morainal debris for several miles above their mouths. Conspicuous terracelike lateral moraines extend along the valley sides. Especially on the north flank of the range in such canyons as Smith Fork, Henrys Fork, and the West Fork of Beaver Creek, the drainage through the morainal belts is interrupted by ponds, marshes, and meadows—an ideal habitat for America's (and no doubt the world's) southernmost moose herd.

In the eastern part of the High Uintas, some valleys, such as Sheep Creek, Carter Creek, and Ashley Creek, were glaciated only in their upper reaches. Downstream from the heads of these valleys, broad V-shaped canyons carved by glaciers give way to extremely narrow precipitous gorges cut entirely by running water. Such gorges are greatly influenced in form and character by the particular rock formation into

Aesop Lake on the East Fork of the Bear River, a glaciated valley in the High Uintas. Pyramidal peaks, hanging tributaries, and flat valley bottoms are scenic products of the Ice Age. (Fig. 8)

which they are cut. For example, canyons eroded into the Weber Sandstone—one of the prime cliff-forming units in the Uinta Mountains—are wild and picturesque. Sheep Creek, Brush Creek (fig. 9), and Dry Fork of Ashley Creek (fig. 44), as well as the wonderful canyon of the Yampa River in the Eastern Uinta Mountains, owe most of their grandeur to the Weber Sandstone. (See also fig. 24.)

A glance at a topographic map shows plainly that the crestline of the Western Uinta Mountains is closer to the north flank of the range than to the south flank. Streams draining the south flank are therefore longer than those draining the north flank. On the average, they are about twice as long. Their canyons are correspondingly long, and the amphitheaters at their heads are both longer and wider. Because the area available for the catchment of snow was larger on the south flank than on

Brush Creek Gorge. A precipitous nonglacial gorge carved by running water. Walls of Weber Sandstone are capped by the Park City Formation. Flat-topped Taylor Mountain in distance is a remnant of the Gilbert Peak erosion surface, an ancient high-level plain. (Fig. 9)

the north, the glaciers there were generally larger, too, except in an early piedmont (Little Dry Glaciation) stage of the north flank, when coalesced sheets of ice spread out beyond the mountain front.

At their maximum the glaciers on the south flank of the Uintas were able to eliminate many of the intervening divides of their headward tributaries and, in so doing, to carve out the great compound cirques, or amphitheaters, which now make the scenery so striking. In the southwestern part of the range, centered near Bald Mountain, the ice succeeded in removing all but a few isolated peaks. These peaks rose as *nunataks,* or islands of rock, above a massive ice cap from which long tongues of ice extended radially down the canyons of the Bear, the Weber, the Provo, the Duchesne, and Rock Creek (fig. 10). Even Bald Mountain and other nearby peaks may have been covered by ice at an early stage of glaciation. Bald Mountain appears to have been reshaped by ice moving across it from the northwest. If the ice ever was thick enough to cover Bald Mountain, the whole western part of the range at one time must have been locked in an immense, unbroken expanse of ice.

The glaciers of the south flank carved much larger amphitheaters than their northerly counterparts did and, thus, were evidently more effective in attacking their headwalls and broadening their cirques—perhaps because they generally were larger and had larger catchment areas. Why, then, should the northerly landscape be more rugged and alpine? The Uinta Mountains are at their scenic best north of the crestline between the Bear River on the west and Henrys Fork on the east; Red Castle at the head of Smith Fork is the crown jewel of the range.

Undoubtedly, the great amphitheaters date from an early glacial episode, perhaps the Little Dry Glaciation, which may really be a composite of several early glacial advances and a time or times when great thicknesses of ice nearly buried the range. The Little Dry ice may thus have opened the cirques, sapped away the tributary divides, and excavated the amphitheaters. Then, after a long warm interglacial period, when weathering had softened the contours of the land and the rivers had cut new canyons into the valley bottoms, the younger Bull Lake and Pinedale glaciers reoccupied the area. Although large, these glaciers were thinner and less massive than their predecessors. They reexcavated the valleys,

Duchesne–Bear River Divide, Western Uinta Mountains. All the dark forested area was covered by a Pleistocene ice cap, which flowed radially outward from the mountains. Low round summits in middle distance were overridden by the ice, as, perhaps, was Bald Mountain (lower left) at an early glacial stage. Photograph by Hal Rumel. (Fig. 10)

eroded new U-shaped inner gorges, and attacked north-facing slopes and hillsides with great vigor, but they had minimal effects on the south-facing slopes of the divides themselves. Almost without exception, the younger glaciers were more effective against north-facing exposures than against south-facing ones.

Rock-Rubble Deposits at High Altitudes

The *felsenmeers* of the Western Uinta Mountains (p. 13) are residual deposits that accumulated in place through long periods of sustained severe climate. Other deposits of loose rock, also riven by frost but transported some distance largely by gravity, have accumulated near the bottoms of slopes at many places in the higher parts of the range. Three distinct kinds of loose-rock accumulations grade into one another: talus, protalus, and rock glaciers. Although they are relatively minor features of the alpine landscape, in bulk and in topographic relief, they contribute greatly to the overall scenery. All three require a brittle blocky bedrock source.

Debris-mantled slopes at the head of Blacks Fork of the Green River. Rocky peaks and ridges are slowly being engulfed in their own talus. Note protalus ramparts, outlined by snowbanks (lower right). Photograph by Max D. Crittenden, Jr. (Fig. 11)

Talus (fig. 11) is a sheetlike or conelike deposit familiar to all hikers in the high country. It blankets the feet of steep slopes or cliffs in large accumulations of loose angular rubble often only marginally stable and containing poorly balanced blocks of rock weighing many tons—a fine place for the unwary hiker to turn an ankle. Talus, however, is not confined to high altitudes in the Uinta Mountains; it is abundant below cliffs everywhere, as in Lodore and Red Canyons along the Green River in the Eastern Uinta Mountains.

Protalus and rock glaciers, on the other hand, are restricted to alpine settings in the Western Uintas, largely to areas above timberline. As such they are abundant in the Kings Peak, Wilson Peak, and Red Castle areas, in the highest part of the range.

Protalus accumulates in rampartlike deposits at the foot of semipermanent snowbanks or ice banks. Rocks are dislodged from above. They roll down the snowbank to its base where, when the snow finally melts, they remain behind as low ridges. And inasmuch as thick accumulations of snow or ice obviously are essential to the process, protalus ramparts formed chiefly in the recent geologic past, the "neoglaciation" of the last few thousand years, when such accumulations of snow and ice were prevalent. Some ice accumulations at that time grew into small true glaciers and, thus, provided a link between protalus ramparts and small glacial moraines. No clear-cut distinction exists, in fact, between moraines and protalus ramparts. They share the same climatic and topographic setting, and they grade transitionally into each other. At the foot of Kings Peak, modern talus cones are forming across inactive neoglacial protalus ramparts.

Rock glaciers are abundant in the Kings Peak, Wilson Peak, and Red Castle areas also. Rock glaciers were first identified by S. R. Capps (1910) in the mountains of Alaska, but they have since been widely recognized in high mountains elsewhere. They have been described in a definitive paper by Wahrhaftig and Cox (1959).

Although rock glaciers consist largely of coarse angular rock fragments, they also commonly have cores of ice mixed with boulders, gravel, sand, and silt. Indeed, fed from above by avalanches and streams of talus, they are thought to move by intergranular flowage between ice and rock. Interstitial ice and a climate in which the ice can form are therefore essential for the nourishment and growth of rock glaciers.

Like true glaciers, active rock glaciers in the Uinta Mountains produce rock flour, a milky suspension of pulverized rock that causes turbidity in nearby lakes and streams. Rock flour is formed by the grinding action of rock particles rubbing against one another during glacier flowage. Rock glaciers resemble small true glaciers in shape, size, and mode of flowage (fig. 12). They form lobate mounds a few hundred feet high, a few hundred to a few thousand feet across, and generally less than a mile long. They have steep snoutlike fronts, sinuous longitudinal ridges analogous to medial moraines, and transverse ridges and furrows analogous to crevasses. They are, in brief, among the more intriguing details of the alpine scene.

The Eastern Uinta Mountains

Although the main divide is very well defined in the Western Uinta Mountains, it is very poorly defined in the Eastern Uintas. Because the Eastern Uinta Mountains are more complex structurally than the Western Uintas, and because they have had a more complex Tertiary history, their physiography is more complex, and they lack the grand simplicity of the west half of the range. The main divide has shifted with time from a position that once must have coincided with the crest of the anticline to its present position on the south flank. The crest of the anticline, moreover, passes beneath the valley of Browns Park, which extends deep into the heart of the range and separates the northeast flank from the rest of the range.

The physiography of the Eastern Uinta Mountains is further complicated by the great canyon systems of the Green and Yampa Rivers. These canyons—sources of awe and wonder since the days of the mountain men—stimulated early-day thinking on many geologic problems, most particularly on folding and faulting, which are grandly portrayed, and on the relations of drainage development to the formation and growth of mountains. Speculation as to how the Green River established its course across the Uinta Mountains led Powell to introduce such terms as "superposition" and "antecedence" to identify processes by which streams are able to establish and maintain courses across mountain barriers. Although Powell coined the terms, he alone did not devise the concepts.

Active rock glacier causes turbidity in adjacent glacial lake, east side of Red Castle. Rock glacier is nourished by talus and avalanching. Lake is dammed by a moraine. (Fig. 12)

Major physiographic subdivisions, Eastern Uinta Mountains. (Fig. 13)

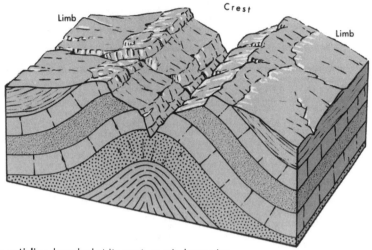

An anticline, breached at its crest, or axis, by erosion.

The canyons of the Green and Yampa are the most impressive features of the Eastern Uinta Mountains, but even without them, the mountains would possess much to fire the interest of the traveler and the imagination of the geologist. Physiographically, the Eastern Uinta Mountains are almost totally unlike their western counterpart. The lofty summits are lacking; no part has been glaciated; and there is no climatic timberline, although many of the summits are bare of trees. The highest point, Diamond Peak—altitude 9,710 feet, and site of the infamous diamond hoax of 1872—is an outlier completely separated from the rest of the range. Mount Lena, south of Flaming Gorge Dam, at an altitude of 9,600 feet, is the highest point on the main divide. Most of the country lies below 8,000 feet.

Lacking high peaks to intercept moisture-laden air masses and standing in the rain shadow of the higher mountains to the west, the Eastern Uintas are relatively arid. Whereas parts of the Western Uintas probably receive 50 inches or more of moisture per year, only the Mount Lena area in the Eastern Uintas is likely to get half that much, and such places as Lucerne and Ashley Valleys get only 8–10 inches. Consequently, the pine, fir, and spruce forests of the Western Uintas give way largely to pygmy forests of juniper and pinyon, although stands of lodgepole pine are dense in the Mount Lena area, and, locally, elsewhere there are groves of ponderosa pine or stands of Douglas fir. The lush high-altitude meadows of the Western Uintas have no eastern counterpart. If any one characteristic thus identifies the Eastern Uinta Mountains, thanks to the aridity of the climate, it is well-exposed bedrock. Bedrock is everywhere—in the bottoms and sides of gullies, along the ridges, in tremendous cliffs along the canyon walls, and on the faces and barren tops of the mesas. The plant cover is light, and the soil is thin.

Topographically, the Eastern Uinta Mountains are divisible into four large highland blocks separated by the canyons of the Green River, the valley of Browns Park, and the canyon of the Yampa River. These blocks, for convenience, can be referred to as (1) the Dutch John–Cold Spring highland, (2) the Diamond Mountain highland, (3) the Douglas Mountain highland, and (4) the Blue Mountain highland (fig. 13). The intervening canyons are formidable barriers that make vehicular access from one highland block to another difficult, if not impossible. Thus, Diamond Mountain, Douglas Mountain, and Blue Mountain are mutually exclusive; none can be reached directly from either of the others. The Dutch John–Cold Spring highland can be reached from the south by a scenic mountain highway (Utah State Highway 44) that connects Dutch John to Vernal, by way of Flaming Gorge Dam. But before the dam was built, access from the south involved many circuitous miles of difficult driving.

Dutch John–Cold Spring Highland

The Dutch John–Cold Spring highland is not really a topographic entity; rather, it is a heterogeneous belt of bare-topped hills, hogbacks, and mesas bounded vaguely by the Green River Basin on the north and the Flaming Gorge–Red Canyon–Browns Park valley complex on the south. On the east it terminates at Vermilion Creek, in a narrow slot 600 feet deep. This highland contains some of the most diverse topography, some of the most complex geology, and the oldest rocks in the entire Uinta Mountains; it also contains Utah's oldest commercial gas field, Clay Basin.

Large flat-topped mesas, such as Bare Top (Bear Mountain), Goslin, and Cold Spring Mountain, are remnants of a broad erosional plain that once extended along the flank of the mountains for the length of the range, between the high peaks on the south and the basin on the north. Named the Gilbert Peak erosion surface and described by W. H.

Bradley (1936), this plain has been faulted and tilted and has been dissected by deep canyons since it was formed.

Most of the Dutch John–Cold Spring highland is on the northeast limb of the Uinta anticline, where a great thickness of upturned strata is beveled by erosion and exposed to view. More than 20,000 feet of strata is exposed in one continuous section on Cold Spring Mountain, even though much of the total section is omitted from surface exposure by faulting. Altogether, the area contains more than 40,000 feet of strata, not counting the ancient metamorphosed Red Creek Quartzite, which forms the crystalline core of the range and which itself is probably 20,000 feet thick. Thanks to large-scale faulting and deep erosion, the Red Creek Quartzite is well exposed along the north side of Browns Park. The age of the Red Creek, as determined by radiometric dating, is 2.3 billion years.

Diamond Mountain Highland

The Diamond Mountain highland, south of Red Canyon and Browns Park, is a dissected upland along the main divide of the Uinta Mountains. As seen from the south near Vernal, it presents an even skyline that belies a more rugged interior. It is surmounted by many craggy peaks and ledges of somber red quartzite and is flanked on the south by a thick sequence of colorful, well-exposed rock formations made up mostly of sandstone and limestone. It is almost completely enclosed by deep canyons and steep escarpments. On the west it is separated from the Western Uinta Mountains by a low pass which also provides the route for Utah State Highway 44. The pass drains north into Cart Creek, which meanders through a verdant meadow and plunges into a gorge a thousand feet deep before joining the Green River near Flaming Gorge Dam. The south side of the pass drains into Little Brush Creek through another gorge nearly as deep and even more precipitous.

On the north the Diamond Mountain highland is bounded by rugged terrain along Red Canyon and Browns Park. Several small spring-fed streams flow north in open valleys with floors of the soft Browns Park Formation. None of these streams has a mean flow of more than a few cubic feet per second. They then descend into narrow quartzite canyons, some of which rival Red Canyon itself, before joining the Green River.

Crouse Creek is typical: its canyon affords access into Browns Park by way of a rough but generally passable truck road. The creek bottom is chocked with brush and trees—cottonwood, boxelder, alder, choke cherry, red-osier dogwood, willow, wild rose, and poison ivy—a good habitat for beaver, bobcat, and cougar.

The relation of these north-flowing streams to Pot Creek—the "Summit Valley" of Powell's day—is very strange. Pot Creek flows southeast the full length of the highland to Lodore Canyon. Yet the several divides between Pot Creek and the north-flowing drainages are low saddles only a few feet high. The whole drainage system of the Diamond Mountain highland is on the Browns Park Formation, which, in turn, fills an older pre-Browns Park valley system cut into the hard quartzite of the Uinta Mountain Group. In many places the present drainage is directly opposite the original drainage, and in some places, creeks flowing in opposite directions share the same pre-Browns Park valley. Many minor tributaries are "barbed"; they flow south into creeks that flow north. The distribution of the Browns Park Formation is very significant because the discordant relation of the Green and Yampa Rivers to the Uinta Mountains is tied to the deposition and removal of the Browns Park Formation.

On the east the Diamond Mountain highland is bounded by the deep chasms of Lodore and Whirlpool Canyons. From the rim of Wild Mountain at the southeast corner of the highland to the river below is a drop of more than 3,000 feet. Though a delight to white water boatmen, these canyons are insurmountable barriers to cross-country travel; west of the canyons, a corner of Colorado is completely cut off from the rest of the state and can be reached only through Utah.

On the south the highland is bounded by the rim of the Diamond Mountain Plateau, a broad nearly flat upland and a likely correlative of the Gilbert Peak erosion surface of the north flank of the range. Diamond Mountain is capped by gravel and volcanic ash of the Browns Park Formation. Diamond Gulch drains southeastward across this area in a course nearly parallel to that of Pot Creek. It joins the Green River in Whirlpool Canyon via Jones Hole, a hidden spring-fed valley of rare beauty.

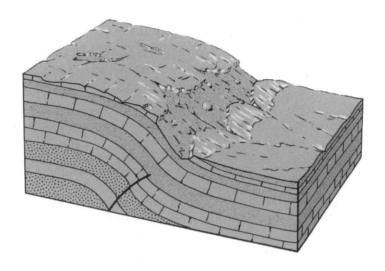

A monocline, faulted at depth.

Douglas Mountain Highland

East of the Diamond Mountain highland, across Lodore Canyon, is Douglas Mountain. This dissected highland, nearly triangular in outline, is a jumble of cliffs, ledges, ravines, and parklike clearings. It is bounded on the north by an arid extension of Browns Park, on the east by the Little Snake River, and on the south by the twisty canyon of the Yampa River. Access is only from the north over rough mountain roads and trails. Zenobia Peak, at an altitude of 9,006 feet, the highest summit, commands impressive views in all directions. It also is the highest point in Dinosaur National Monument.

Douglas Mountain is an area of brightly colored rocks. The brilliant white exposures of the Browns Park Formation, dazzling in the sun, contrast with the deep-red hues of the adjacent Uinta Mountain Group. Blue, gray, and pale-pink limestones cap Zenobia Peak and reach south in long sloping ridges toward the Yampa River. The canyons themselves are variegated in reds, gray, and buff, depending partly on the particular rock formation exposed in the walls and partly on the time of day. Douglas Mountain has a few small springs but no perennial drainage.

Blue Mountain Highland

South of Douglas Mountain, across Yampa Canyon, is Blue Mountain, the southeasternmost highland block of the Uinta Mountains. Steep, even precipitous, escarpments on the west, south, and east of this highland block are surmounted by an open, rolling interior, which rises gradually northward to the bald summits of Round Mountain, Marthas Peak, and Tanks Peak. Rewarding views are had from these peaks, also. To the north, the ground drops away more than 2,000 feet in a giant step down to the flats that rim Yampa Canyon. The river itself is another 1,000 feet below.

Abrupt monoclinal folds characterize the flanks of the Blue Mountain block, and they are grandly displayed. Monoclines are steplike bends in otherwise flat or gently dipping strata. The monoclines were formed by the warping of strata across concealed faults, or by the sharp flexing of strata on the flanks of flat-topped anticlines. On Blue Mountain these structures owe their dramatic appearance to the erosive resistance of the Park City Formation, to the massive character of the underlying Weber Sandstone, and to the easy erosion of the overlying Moenkopi Formation. As the soft Moenkopi is stripped away, the Park City and Weber are left standing in enormous dip slopes and "flatirons."

One of the more spectacular folds of this type is just north of U.S. Highway 40, where Stuntz Ridge, a westward extension of Blue Mountain, forms a jutting promontory nearly 3,000 feet high (fig. 14). This remarkable flexure passes downward and laterally into a large fault.

Other flexures nearly as dramatic form the north face of Blue Mountain, where the strata have been folded into sharp monoclinal bends along the Yampa fault and subordinate fractures in the stairstep noted before. These structures are seen to advantage from observation points along the high road to Harpers Corner, from Harpers Corner itself, or from the access road to Echo Park far below (fig. 26). Harpers Corner, incidentally, is the best canyon overlook in Dinosaur National Monument. It projects north as a sharp ridge from the main mass of Blue Mountain, where it looks down on Whirlpool, Lodore, and Yampa Canyons, and on Steamboat Rock at the confluence of the Green and Yampa Rivers.

Stuntz Ridge, a sharp monoclinal bend on the south flank of Blue Mountain. Eroded "flatirons" below the ridge were once continuous with strata on top. (Fig. 14)

To the west, on the same structural trend as the Yampa fault, is Split Mountain, a nearly symmetrical anticline breached to its core by the Green River (fig. 29). The outer shell of Split Mountain consists mostly of Weber Sandstone, which is deeply eroded into parapets, buttresses, and other fanciful architectural forms; even so, the outer shell reflects the overall shape of the fold. Younger rocks at the flanks form encircling hogback ridges and racetrack valleys; hard layers are etched into relief. The famous Dinosaur Quarry is on the south limb of the fold, where the dinosaur-bearing Morrison Formation is laid bare by erosion.

Split Mountain has been a source of wonderment since the days of Powell, particularly because a seemingly easier course for the Green River is available in the soft beds of the racetracks that encircle the anticline, as opposed to the hard rocks in the core. The river, however, had no other options at the time that downcutting began. Split Mountain did not exist as a topographic feature, and once entrenched, the river had no alternative but to pursue its established course. The "easy" course in the racetrack was eroded out later by tributaries. Meanwhile, the shape of the fold as brought into relief as soft beds were removed and hard beds were uncovered.

The Great Canyons of the Green and Yampa Rivers

The Green and Yampa Rivers arise in distant watersheds, fed by snowmelt from the Wind River and Park Ranges, as well as from the Uintas. Both streams flow in broad valleys, but they enter the Uinta Mountains in narrow gorges and flow across the mountains with a startling disregard for the geologic implications of their crossing. The sequence, altitudes, and gradients of canyons through the range are shown graphically in figure 15.

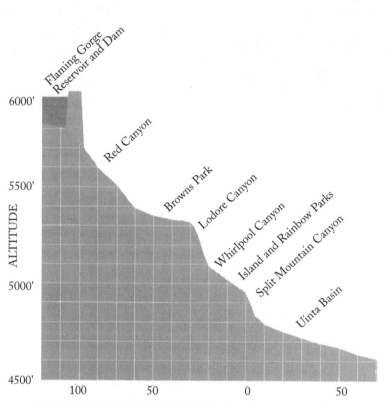

Graphic profile of the Green River across the Uinta Mountains from Flaming Gorge Dam to the Uinta Basin. Data from Woolley, 1930, pl. 30. (Fig. 15)

The Geologic Story of the Uinta Mountains

Flaming Gorge, viewed downstream, as photographed by E. C. LaRue in 1914. Flaming Gorge Reservoir now fills the lower third of the canyon. Glen Canyon Sandstone, topping the cliff, overlies Chinle and Moenkopi Formations. (Fig. 16)

One hundred years ago Powell and his aides floated leisurely through the twisty canyons of southern Wyoming on the first leg of their historic boat trip down the Green and Colorado Rivers. In Powell's time the Green was regarded as the source of the Colorado. Just below the 41st parallel—now the Utah state line—the current quickened, and Powell's party abruptly entered the Uinta Mountains. "The Green River enters the Uinta Mountains by a flaring, brilliant, vermilion gorge, a conspicuous and well-known locality, to which, several years ago, I gave the name Flaming Gorge" (Powell, 1876, p. 146). The same scene, viewed by E. C. LaRue in 1914, is shown in figure 16.

Flaming Gorge to Browns Park

Flaming Gorge Dam, completed in 1964, has greatly changed the appearance of the country upstream. Today, a man-made lake extends 90-odd miles to Green River, Wyoming; it forms a broad embayment above Flaming Gorge but constricts tightly at the gorge entry. The rapids are quiet, replaced by a fjordlike body of still, blue water, hardly more than twice the width of the former river in some places but elsewhere widened into bays and inlets. Whole mountains are now islands.

Flaming Gorge is cut into Jurassic and Triassic rocks. Its sweep and grandeur are due largely to its high cap of Glen Canyon Sandstone (the Nugget or Navajo Sandstone of earlier reports), but it takes its name from the riot of reds, ochers, and oranges in the shaly Triassic beds below. It is more a water gap than a gorge in the usual sense. At the portal the rocks are standing on edge, but, arching over, they flatten rapidly a short distance downstream. Just below Flaming Gorge, the shoreline of the reservoir swings left into the Weber Sandstone, which forms the smooth walls of Horseshoe Canyon. Curving right, through Horseshoe Canyon, the reservoir then turns back into Triassic rocks in a broad curve to the left, then right again into the Weber in Kingfisher Canyon. Powell met his first real rapid at the mouth of Kingfisher Canyon, where he crossed the great Uinta fault and entered Red Canyon (fig. 17). Red Canyon, about 30 miles long, is cut entirely in Precambrian rocks, except at Little Hole, a parklike opening underlain by the Browns Park Formation and a pleasant contrast to the rugged canyons upstream and down.

The roar of the river overprints the scenery of Red Canyon. Red Creek Rapid below Little Hole is particularly unrestrained. Just about all the rapids in Red Canyon are at the mouths of tributaries, where from time to time, coarse rock debris is flushed into the main channel by flash floods. Most rapids in Lodore, Yampa, and Split Mountain Canyons were formed in the same way.

At the mouth of Red Canyon the river emerges into Browns Park, the legendary hideout of Butch Cassidy and his notorious "Wild Bunch," who terrorized southern Wyoming, eastern Utah, and western Colorado near the turn of the 20th century.

Red Canyon, about 2½ miles below Flaming Gorge Dam. Red Canyon is cut in flat-lying Precambrian quartzite (Uinta Mountain Group). (Fig. 17)

Browns Park

Browns Park is a picturesque intermontane valley, which, in Powell's words, is "really an expansion of the cañon." It is broad, averaging about 4 miles across, but it is much longer than wide. Its actual length is arguable, but most people believe that it extends from Red Creek on the west to Vermilion Creek on the east, a length of about 25 miles.

Steep mountain fronts bound the valley on both sides, the mountaintops rising 3,000 feet or more above the valley floor. The north front is more abrupt than the south, partly because it is bounded for most of its length by faults, which drop the valley side relative to the mountains. Small creeks, after descending from the heights, enter the valley from both sides through deep brushy canyons.

Browns Park gains most of its distinctive appearance from its thick fill of Tertiary gravel, sand, clay, and volcanic ash—the Browns Park Formation. This fill forms the broad flat bottom of the valley, and its brilliant white exposure contrasts with the somber appearance of the nearby mountains (fig. 18). Flat-topped Quaternary terraces are characteristic also.

Browns Park is doubly curious because of its structure. The Browns Park Formation accumulated on the floor of a preexisting canyonlike valley. The valley was formed partly by ordinary erosion and partly by downfaulting along the crest of the great Uinta anticline. Faulting probably helped to establish drainage there in the first place. As the fill accumulated, the contour of the old valley gradually softened, until it took the shape of a shallow elongate syncline (something like an extra-long celery dish). Thus, we have a syncline of deposition, modified by later faulting, superimposed on the crest of a folded anticline. The axis of the syncline is well north of the present valley bottom, which crowds the south margin of the valley. Clearly, the deepest, thickest part of the old fill is well to the north of the present river (fig. 19).

As the old fill spread over the floor of Browns Park, it slowly rose up the valley sides until it eventually buried tributary canyons and intervening ridges alike. Ultimately, it probably filled the valley to the rims. Then, when downcutting was resumed, the soft fill was removed preferentially, so that buried promontories, such as Kings Point, were exhumed—an ancient Tertiary landscape faithfully, if incompletely, restored. Tertiary drainage through Browns Park was obstructed from

Browns Park viewed north toward O-Wi-Yu-Kuts Mountain. Steep mountain front is bounded by faults. The white Browns Park Formation contrasts sharply with the somber Uinta Mountain Group. Flat top of O-Wi-Yu-Kuts Mountain is a remnant of the Gilbert Peak erosion surface, a broad Tertiary plain which once flanked the Uinta Mountains. (Fig. 18)

time to time by one thing or another: alluvial fans of sand and gravel spread across the valley from its margins, chiefly its north margin, and formed lakes. Earth movements along faults may also have caused impoundments. Tiny organisms, such as ostracodes and diatoms, thrived in standing water, and their remains were preserved in the lake-bottom muds. The environment may have been too hostile for more complex forms of life, for no other remains have been found.

Frequent falls of volcanic ash blanketed the whole countryside. Ash that fell in the valley is preserved just as it accumulated; it gives the Browns Park Formation its white brilliance. Much of the ash that fell on the adjacent hills was washed into the valley by rains and freshets. Thus, the resulting Browns Park Formation is a varied accumulation of stream-laid sand and gravel, lacustrine clay, and volcanic ash. Its known thickness is more than 1,500 feet.

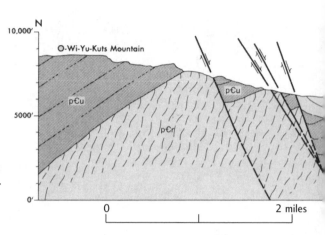

Geologic section across Browns Park, showing troughlike form of Browns Park syncline superimposed on Uinta anticline and showing normal faulting along north valley margin. Browns Park is not notably faulted on the south, although some investigators have suggested that it is. pɛr, Red Creek Quartzite; pɛu, Uinta Mountain Group; Tbp, Browns Park Formation. (Fig. 19)

After entering Browns Park, the Green River flows swiftly but quietly for several miles across the Browns Park Formation. Just after crossing the axis of the Uinta anticline on the south side of the valley, the Green enters Swallow Canyon (a short but steep-walled gorge) cut into the Uinta Mountain Group in a classic example of drainage superposition. The course of the river through Swallow Canyon was fixed at a higher altitude than now in soft Browns Park beds that have since been removed. As the river eroded downward, it reached a buried spur of hard red quartzite through which it was constrained to cut its gorge. The spur, since exhumed, is now Kings Point, high above the canyon walls. Although Swallow Canyon is perhaps the most obvious example of superimposed drainage in the entire Uinta Mountains, many other canyons in the region—and elsewhere in the Rockies for that matter—were carved in the same way. Below Swallow Canyon the river swings back again onto the Browns Park Formation, and as its grade flattens and its current slackens, it meanders lazily the 20 miles or so to the Gates of Lodore.

Lodore Canyon

As noted before, the river crosses the axis of the Uinta anticline in Browns Park above Swallow Canyon. Lodore Canyon, therefore, is on the south limb of the fold. The river enters Lodore Canyon deep in the red Precambrian quartzite of the Uinta Mountain Group. And, inasmuch as the dip of the strata, though gentle, is steeper than the gradient

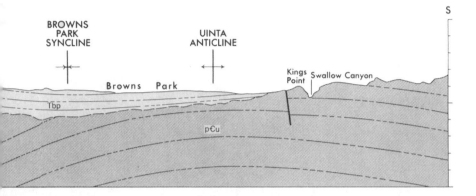

Lodore Canyon, viewed north toward the Gates of Lodore and Browns Park, from Douglas Mountain. The west wall is 2,280 feet high. (Fig. 20)

of the river, the river crosses successively younger rocks en route. About halfway through the canyon, above Triplet Falls, the Lodore Formation (Cambrian Period), the Lodgepole Limestone, and the Deseret Limestone (both Mississippian) first come into view. The Mississippian limestones form a sheer gray cliff 400 feet high that surmounts the somber red beds below. At each bend of the river the cliff descends lower, finally reaching river level 2 miles above the mouth of the canyon. At the mouth of Lodore Canyon, at Echo Park, the river flows on Weber Sandstone, having crossed strata, about 1½ miles thick, of Precambrian, Cambrian, Mississippian, and Pennsylvanian age. Then, around the next bend, a sharp fault (Mitten Park fault) has raised the Precambrian back again above river level.

Lodore is the deepest and most precipitous of the great river canyons in the Uinta Mountains (fig. 21), and the Gates of Lodore could hardly be more imposing. Looking into the gates, one senses some of the mixed emotions—exhilaration, wonder, and apprehension—that must have bestirred Powell and his boatmen.

The river, after meandering parallel to the mountain front through Browns Park, swings to the right in a broad arc and drives headlong into the canyon. The current quickens, but the river flows smoothly and in a nearly straight line for about 3½ miles. Then, a thundering roar ahead signals approaching turbulence, and after a sharp bend to the right, the river plunges into its first rapid. It drops from one rapid into another for the next 8 miles, with short respites of quiet water in between. Powell's names for some of these rapids—Upper and Lower Disaster Falls, Hells Half Mile—give some idea of the fury of the river.

At the Gates of Lodore the canyon walls rise abruptly from the water's edge. A mile downstream the canyon rim is more than 2,000 feet above the river. The walls are most precipitous near the head of the canyon, although they are higher downstream at Wild Mountain, where the drop from rim to river is more than 3,000 feet. Lodore Canyon may have been cut in two stages. As shown in figure 22, a steep inner gorge rises 600–800 feet above the river, then a series of buttresslike forms slope up several hundred feet to the base of precipitous outer walls a thousand

West wall of Lodore Canyon, as photographed by James Gilluly in 1921. Buttresslike forms halfway up canyon wall suggest two stages of canyon cutting—an inner gorge incised into an outer one. (Fig. 21)

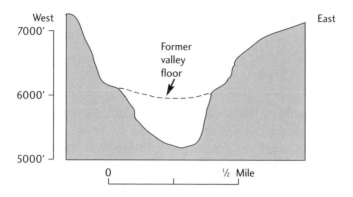

True-scale profile of Lodore Canyon, just below Gates of Lodore, suggesting two stages of canyon cutting. Canyon was about 1,500 feet deep when postulated reentrenchment began. Compare with figure 21. (Fig. 22)

feet higher. A canyon profile (fig. 22) suggests that the postulated outer canyon was about 1,500 feet deep when the inner gorge was cut, probably in Quaternary time.

Yampa Canyon

At Echo Park, or Pats Hole, as it is also known, the Yampa River flows into the Green. Echo Park (fig. 23) is a longtime favorite of canyon visitors and is accessible to automobiles by way of a rough but scenic dirt road. Yampa Canyon extends upstream about 45 river miles from Echo Park to a point near the east end of the Uinta Mountains called Lily Park. Just below Lily Park the Yampa River cuts abruptly across Cretaceous, Jurassic, Triassic, and Permian strata, then across the Weber Sandstone, and into older Pennsylvanian rocks generally called the Morgan Formation (fig. 24). The river flows across the Morgan at about the same stratigraphic level for 18 miles. It then cuts back into the Weber Sandstone and stays mainly in the Weber to its confluence with the Green.

Because of marked differences in color, bedding habit, and lithology between the Morgan and the Weber, the appearance of Yampa Canyon changes drastically across the formation boundary. Through the Morgan, the canyon is mostly wide and asymmetrical; it is steep, or precipitous, on the south side, and ledgy and sloping on the north. Layers

Echo Park (lower right) and Yampa Canyon, carved from massive Weber Sandstone, which dips gently to the south (right). Viewed east from Harpers Corner. Steamboat Rock in foreground. National Park Service photograph. (Fig. 23)

of pink sandstone and siltstone alternate with gray, tan, and pink limestone. The course of the river is curvy but not tortuous. Through the Weber Sandstone the canyon is narrow, precipitous, and spectacular. Smooth canyon walls overhang the river in places, rising sheer for hundreds of feet above the water, particularly on the outside bends of meanders. The course of the river is extremely twisty—in less than 10 miles, straight line, the river meanders 22 miles, and each turn opens a new vista.

Yampa Canyon. Weber Sandstone (cliff in background) overlying the Morgan Formation. National Park Service photograph. (Fig. 24)

Whirlpool Canyon and Island Park

Just below Echo Park the Green River makes a long hairpin bend around Steamboat Rock, crosses the Mitten Park fault, and enters Whirlpool Canyon, a spacious gorge about 2,500 feet deep and 1½ miles across. Shown in profile in figure 25, the walls of Whirlpool Canyon are less precipitous than those of Lodore Canyon and are less picturesque than those of Yampa, but they are very impressive, nevertheless. The dip of the strata, again, is down stream, so that the river flows across successively younger rocks—the same formations exposed in Lodore Canyon. The Uinta Mountain Group forms a vertically walled inner gorge 300 feet deep in the headward part of the canyon, and the river fills the bottom, wall to wall. Then at the first bend, the Uinta Mountain Group passes below drainage. Although the heights of the canyon are eroded mainly from Pennsylvanian and Mississippian strata, Cambrian rocks (Lodore Formation) are well exposed in the canyon bottom from the Mitten Park

Whirlpool Canyon, viewed downstream from Harpers Corner. Stair-steps of cliffs and ledges are caused by resistant and nonresistant strata. (Fig. 25)

fault downstream to about mid-canyon. In the upper canyon walls, massive limestone beds of Pennsylvanian and Mississippian age form giant stair-steps of sheer cliffs and sloping ledges, capping off the rims at Wild Mountain and Harpers Corner.

Halfway through Whirlpool Canyon, entering from the right (north) is spring-fed Jones Hole Creek, a sparkly stream that issues at full volume from openings in the Pennsylvanian Round Valley Limestone. Jones Hole is dominated by towering cliffs of Weber Sandstone in a picturesque and dramatic setting. It is one of the foremost attractions of Dinosaur National Monument and a remembered highlight of any float trip down the Green. It can also be reached overland, on foot, from the Jones Hole National Fish Hatchery.

Mitten Park fault as seen from Harpers Corner. Photograph by Dale Thompson, National Park Service. (Fig. 26)

The Geologic Story of the Uinta Mountains

Island Park, viewed northeast. Island Park fault separates uplifted block (right) from downfaulted block (left). Photograph by Clint McKnight. (Fig. 27)

Large faults bound Whirlpool Canyon upstream and down—the Mitten Park and Island Park faults, respectively. Between these faults, the canyon area has been raised about 3,000 feet relative to the areas upstream and down. One must realize, however, that the faulting took place long before the canyon was eroded. The Mitten Park fault can be seen very well from the river and from Harpers Corner high above (fig. 26). The Island Park fault (fig. 27) can be seen well from the mouth of Whirlpool Canyon and in Jones Hole.

Whirlpool Canyon opens downstream through a flaring gateway into Island Park, which together with Rainbow Park is an attractive river valley of a few square miles between Whirlpool and Split Mountain Canyons. Island Park is on the axis of a faulted asymmetrical syncline, first cited by Powell; it trends west between the main Uinta anticline to the north and the Split Mountain anticline to the south (fig. 28). The syncline axis is crowded against the flank of Split Mountain. Cretaceous rocks are preserved in the trough of the syncline, their first occurrence downstream from Flaming Gorge.

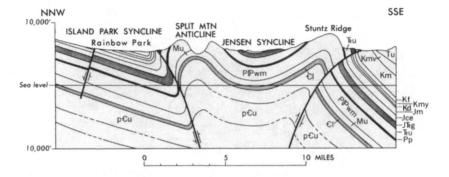

Geologic section through Split Mountain, showing inferred relation of faults to folds. Vertical scale is exaggerated x 2. p€u, Uinta Mountain Group; €l, Lodore Formation; Mu, Mississippian rocks; PₚPwm, Weber and Morgan Formations; Pp, Park City Formation; ₮Ru, Triassic rocks; J₮Rg, Glen Canyon Sandstone; Jce, Curtis, Entrada, and Carmel Formations; Jm, Morrison Formation; Kd, Dakota Sandstone; Kmy, Mowry Shale; Kf, Frontier Formation; Km, Mancos Shale; Kmv, Mesaverde Group; Tu, Tertiary rocks. (Fig. 28)

Split Mountain Canyon

After meandering placidly through Island and Rainbow Parks, the Green River heads into Split Mountain Canyon, an eroded anticline (fig. 29). Flowing swiftly across upturned beds, the river passes from Jurassic rocks at Rainbow Park onto rocks of Mississippian age in the core of the Split Mountain anticline. After crossing the fold axis, the river then recrosses the same beds, but in reverse order, on the south limb of the fold. Varicolored Mississippian, Pennsylvanian, and Permian rocks form the sloping inner canyon walls. Weber Sandstone forms the heights, eroded into a craggy terrain of fanciful turrets, domes, and ramparts. On the south flank of the Split Mountain anticline, the Weber is eroded into an array of buttresslike forms—great monoliths separated from one another by deep, extremely narrow tributary ravines, some of which expand upstream into wide but nearly hidden alcoves.

The Green River runs fast through Split Mountain Canyon, having there its greatest average rate of fall—20.7 feet per mile—in the Eastern Uinta Mountains. At the mouth of the canyon, the gradient flattens, the current slackens, and the river emerges at last from the Uinta Mountains into the Uinta Basin, 118 miles below its point of entry at Flaming Gorge.

How the Canyons Were Formed

Catastrophism as a rationale to explain the origin of canyons has gone out of style. In the past, many people viewing the great river gorges of the west would have guessed that the earth just split open, sundered by some frightful cataclysm, the waters then simply plunging headlong into the void. Cataclysmic events have indeed been recorded in geology. The overflow of extinct Lake Bonneville was such an event; so was the Alaska earthquake of 1964. But these events are rare and, anyway, are not on the order of magnitude needed to cut canyons across mountain ranges. The energy requirements of canyon cutting are infinitely greater, although the energy is expended much more slowly. So, the concept that the great canyons were carved slowly and inexorably by their own drainage is a truism that most people now accept without further explanation.

In fact, the events leading up to the cutting of a canyon by a river can be quite complex and can involve the interaction of many separate processes. To be sure, the river is the chief agent of erosion and the prime mover of eroded rock, but it alone could not do the job. Atmospheric weathering—chemical decomposition, as well as mechanical disintegration of rock—frost action, gullying, rill action during showers, and downslope movement of rock debris under the influence of gravity all help prepare the ground, widen the canyon walls, and transport material to the canyon bottom. Only then does the river take over, wearing away and carrying off its suspended load and, at the same time, attacking and eroding its bed.

The river is most effective during high water, in early summer, when runoff is swollen by snowmelt from the high mountains or by thunderstorms over the watershed. As the volume of water increases, so does the velocity. Increase the velocity, and the carrying power increases geometrically. If the velocity is doubled, the capacity is tripled—a principle spelled out many years ago by G. K. Gilbert, whose namesake Gilbert Peak dominates the Western Uinta Mountains on the north. The roily water of flood stage is an obvious expression of the erosive power of the river. And no one who has seen a large river in flood can doubt its ability to do its work.

But the most intriguing question related to the origin of the great canyons of the Uinta Mountains is not so much the mechanics of erosion as it is how the rivers were able to cross the mountains in the first place.

Split Mountain, viewed from the south, showing the nearly symmetrical form of the anticline. Photograph by R. B. Beidleman. (Fig. 29)

How could rivers establish and maintain their courses across a great mountain range in utter disregard for the structural complexities within the range? Such canyons as Swallow and Split Mountain provide clues, but not answers, and the evidence for the whole canyon system is even less obvious. Differences of opinion through the years have marked the development of thought.

Three concepts, at one time or another, have gained some acceptance: antecedence, superposition, and stream capture. Antecedence, the concept favored by Powell, assumes that the drainage pattern predates uplift and that, as uplift begins and folding progresses, vigorous rivers are able to lower their beds as fast as the land is elevated. Antecedence has been authenticated in some drainage basins, but cogent evidence rules it out for the Eastern Uinta Mountains. For example, early in the history of the range, drainage clearly was away from the rising mountain mass, just as one would expect, rather than toward or across it, as it is now. Rock materials eroded from the heights were redeposited at the flanks, where they now comprise several rock formations younger than the mountains but older than the canyons. As the mountains rose, closed basins formed on both flanks. These basins, the Uinta and Green River, contained great lakes whose deposits now make up the Green River Formation, famous for its fossil fishes, oil shale, and enormously large deposits of sodium salts. Finally, broad erosion surfaces or pediments (for example, the Gilbert Peak erosion surface) were formed on both flanks of the range, sloping outward to the basins. As long as these surfaces were intact, drainage across the range was impossible.

Emmons in 1877 propounded the concept of superposition. He believed, and I think correctly, that the Green River established its course on an old fill of Tertiary sediments (the Browns Park Formation) that spread across the eastern part of the range after the range had been uplifted and deeply eroded. Much of this old fill is still preserved. The Green River, then, according to Emmons' concept, cut down through the fill and incised itself into the buried, older rocks in the core of the range. The Green was thus able to cut across anticlines, synclines, and broad fault zones without undue deflections in its course. It was, in fact, compelled to do so because it was held in by its own banks, and the deeper it cut, the more firmly committed it was to its course. The present discordance between drainage and topography is therefore due more to selective erosion of hard and soft rocks after the drainage pattern was fixed than to differential uplift.

W. H. Bradley (1936) studied the north flank of the Uinta Mountains in the 1920s and '30s. Bradley surmised that the ancestral drainage of the Green River Basin was east, toward the North Platte River, and possibly into the Gulf of Mexico. He also found evidence that the ancestral drainage in the Browns Park area was eastward, possibly into the ancestral Yampa River, or possibly across the present Continental Divide and into the Mississippi drainage system as well.

As a topographic eminence, the present Continental Divide is indeed very subdued all the way from the southern Wind River Range to the northern Sierra Madre in southern Wyoming. In that area, broad closed depressions actually exist along the divide, and the true position of the divide is very obscure. The Sweetwater River arises on the west side of the Wind River Range in what structurally is part of the Green River Basin; it turns south, then east, and flows to the North Platte. Thus, even now, part of the drainage from the Green River Basin flows to the Gulf of Mexico.

Bradley's suggestion of eastward drainage is also supported by remnants of an old valley filled with the Browns Park Formation. The old valley trended east from Browns Park toward Craig, Colorado. Beyond Craig, the valley may have extended north into Wyoming toward Baggs, but the evidence is obscure. Northeast of Baggs, remnants of the Browns Park Formation or its equivalent extend well into the present North Platte drainage, though there is no real assurance that any of those deposits originated in the Uinta Mountains region.

Bradley agreed that the course of the Green was superimposed upstream and down from Lodore Canyon, but he argued that if the course were superimposed through Lodore, an inordinate amount of fill had to have been removed from Browns Park. Bradley also called attention to the relatively straight course of Lodore Canyon, which he contrasted with the sweeping meanders farther upstream and down. Following a suggestion by J. D. Sears (1924), Bradley postulated that an east-flowing stream in Browns Park (the ancestral Green) was diverted south by a small but vigorous stream eroding headward in what is now

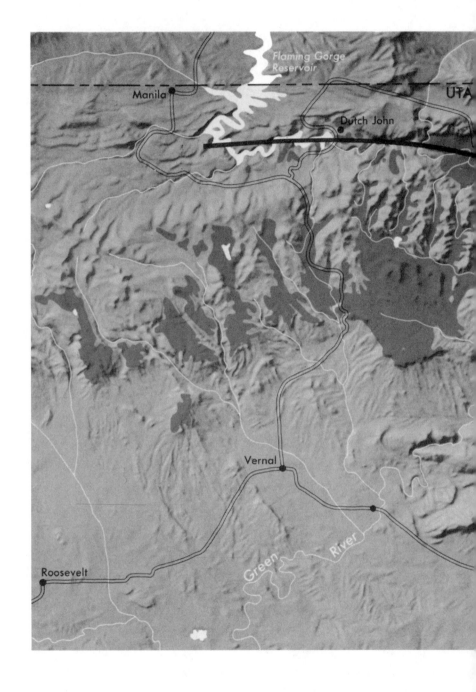

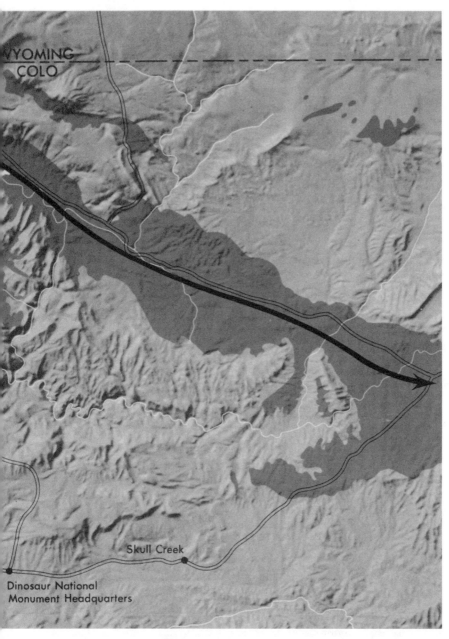

Present distribution of the Browns Park Formation in the Eastern Uinta Mountains. Small unmapped remnants rest on Blue and Douglas Mountains, also. Hypothetical course of the ancestral Green River is shown by the heavy line. (Fig. 30)

Lodore Canyon. This stream, nibbling away at the intervening divide, undercut the channel of the ancestral Green River and turned it southward. Bradley thus introduced the concept of stream capture to explain Lodore Canyon and the southward flow of a river that very clearly once flowed east.

In more recent years, widespread Tertiary fills have been found in the highlands south of Browns Park (fig. 30). These fills are identical with the Browns Park Formation and no doubt correlative with it. Along Pot Creek, Crouse Creek, Cart Creek, and other streams on Diamond Mountain are deposits of white volcanic ash and sandstone at altitudes comparable to the rims of Lodore Canyon. In fact, most of the valleys are blanketed by such deposits; only the ridges and peaks protrude through them. At widely scattered localities on Douglas Mountain, Harpers Corner, and Blue Mountain, deposits of conglomerate and volcanic ash probably are eroded remnants of the Browns Park Formation also. Most of Harpers Corner is capped with gravel or conglomerate. The extent and distribution of all these deposits demand that the old Tertiary fill was very thick indeed and that a tremendous volume was surely removed by erosion. These deposits, therefore, seem to confirm Emmons' view that the course of the Green River across the Uinta Mountains is superimposed.

The establishment of the Green River across the Uintas, as I see it, is thus postulated as follows: An ancient river flowed eastward from Browns Park, as Bradley suggested, hypothetically toward the north end of the Sierra Madre and into the North Platte. If so, the crestline of the Uinta Mountains at that time was the Continental Divide. Steps in the drainage history are shown in figure 31. Meanwhile, crustal movements preceded, perhaps accompanied, and certainly followed deposition of the Browns Park Formation. In brief, the whole eastern part of the Uinta Mountains between the Uinta fault on the north and the Yampa fault on the south was involved in a complicated "graben" movement, or collapse. First noted by Powell, the collapse of the Eastern Uintas has since been affirmed by many geologists, particularly by Sears and by Bradley. The present Continental Divide across central Wyoming is viewed as having begun to rise at about the same time. All these crustal movements, combined with immense outfallings of volcanic ash from some distant source, led to the stagnation and ultimate ponding of all the drainage in the

Eastern Uinta Mountains. The old valley of Browns Park finally became filled to overflowing with a great thickness of gravel, sand, clay, and volcanic ash. The fill spread onto and across the main highland mass of the mountains to the south, leaving only the peaks and ridges exposed. Sears expressed much the same view in 1924. The fill also extended east to west at least from Craig to Flaming Gorge and north to south across the Uinta Mountains from Browns Park at least to Diamond Mountain. It enveloped Cold Spring Mountain on the north and possibly even overtopped it. Its southward limit in the Uinta Basin is unknown, for it has been erased by erosion.

Conditions were then ripe for superposition. With drainage obstructed to the east by the rising Continental Divide and with Browns Park filling with sediment, the stagnating, meandering Green River spilled across the Uinta crest, along what is now Lodore Canyon, to the Uinta Basin, where it became tributary to the Colorado. Below the spillover point, the course of the river probably was nearly straight, just as it is today.

Drainage presumably had already been established out of the Uinta Basin to the south, across the high Tavaputs Plateau. The Green there has since cut a canyon 4,800 feet deep, with a rim 9,200 feet above sea level. Either this rim was breached before the Eastern Uinta Mountains collapsed or it rose after it was breached, possibly both; otherwise the basin would have drained north across the Uintas to the Gulf of Mexico. Even now, the Continental Divide in southern Wyoming is little more than 6,800 feet above sea level—much lower than the rims of the Tavaputs Plateau (and scarcely as high as the rims of the Lodore).

In any event, the Green River, rejuvenated in its new course, must have quickly incised through the soft Browns Park Formation atop the Uintas down into the harder rocks underneath.

At about this time, as thus visualized, the main drainage of the Green River Basin was still flowing eastward at a hydraulic disadvantage across the rising Continental Divide. But the invigorated Green River, with a new, lower base level, now began to entrench its meanders upstream from Browns Park, carving out the spectacular loops in Red Canyon, Horseshoe Canyon, and Flaming Gorge. Near Flaming Gorge the river captured drainage that once flowed north into the Green River

Postulated Development of the Green River Drainage System in Tertiary Time. (Fig. 31)

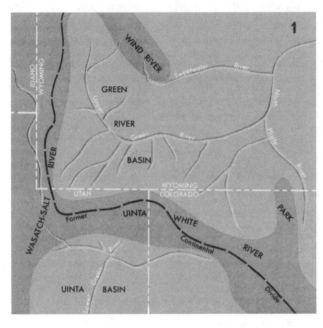

1. Drainage pattern after disappearance of the Green River lakes and shortly after cutting of the Gilbert Peak erosion surface. Drainage is away from mountainous uplifts shown by dark-gray tint. Upper Green River drains east into North Platte. Former Continental Divide is shown by heavy dashed line. Time, possibly earliest Miocene.

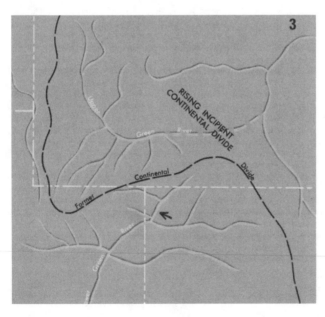

3. Eastward drainage in Uinta Mountains, now flowing on a thick Browns Park fill (not shown), spills across Uinta crest at site of Lodore Canyon (arrow) and begins to incise itself. Eastward flow of Upper Green river is being ponded by rising incipient Continental Divide. Pliocene time.

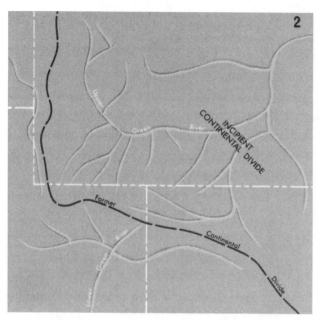

2. Eastward drainage has been initiated in Eastern Uinta Mountains by collapse of Uinta anticline. Collapse continues as Browns Park Formation begins to accumulate (not shown) and as incipient Continental Divide begins to rise. Time, possibly late Miocene.

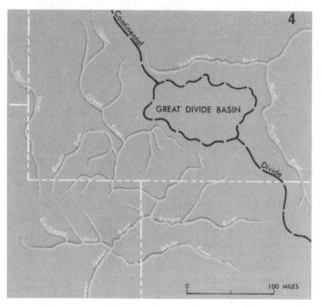

4. Modern drainage. Continental Divide shifts eastward as Upper Green River is captured and turned south by rejuvenated drainage across Uinta Mountains. Canyon cutting continues.

Basin and possibly even reversed the flow direction of the present reach of river between Flaming Gorge and Green River, Wyoming. Presumably, it then captured the main drainage of the Green River Basin. Bradley (1936) noted that the main Uinta Mountain tributaries of the Green flow well north into the basin before turning east and south into the Green, as if in response to a diverted master stream.

Meanwhile, the course of the Yampa River was being established on the south flank of the range. The ancestral history of the Yampa is still in some doubt, even though it has received considerable attention. The pattern of drainage suggests that the Yampa may once have even flowed east from the Uinta Mountains, along with the Green. The Yampa may have joined the ancestral Green somewhere near Cross Mountain. At any rate, the remarkably entrenched meanders of the Yampa in Yampa Canyon point clearly toward superposition of that reach of river. J. D. Sears (1962) discussed in detail the character and origin of these features. Once again, the distribution of the Browns Park Formation leads to the inference that

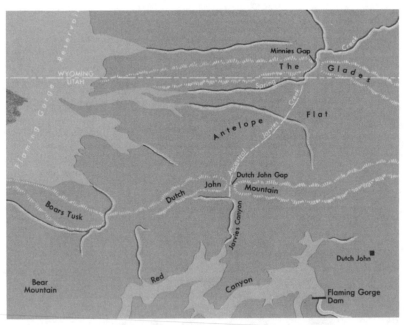

Drainage adjustment at Dutch John Gap. Dutch John Gap was abandoned when ancestral Jarvies Creek, which flowed across the structural "grain," was beheaded by Spring Creek, which eroded headward (east) in the Hilliard Shale. A future capture north of The Glades could cause abandonment of Minnies Gap. Note also imminent drainage diversion just south of Boars Tusk. (Fig. 32)

the Yampa entrenched itself down through a Browns Park fill onto the underlying, harder formations. There, down through hundreds of feet of hard rock, it faithfully reproduced the intricate meander pattern that it had originally developed on an old valley fill long since removed.

Other Drainage Adjustments

Many other streams in the Uinta Mountains have had their courses drastically altered during the past few million years. Particularly at the flanks of the range, drainage became channeled into courses controlled by alternately hard and soft strata. In such areas, as downcutting progressed in belts of tilted rock, streams eroding across resistant ledges tended to yield to streams that flowed along softer strata. Much like a board left to weather in the elements, the harder layers were etched into relief, the softer ones were hollowed out.

Drainage of this sort, said to be structurally adjusted, is clearly visible in the Flaming Gorge area, where several ancestral drainage lines are indicated by abandoned wind gaps. Dutch John Gap is an excellent example; Irish Canyon, at the east end of Cold Spring Mountain, is another (Sears, 1924). Dutch John Gap once channeled the drainage that now flows down Spring Creek; Irish Canyon may once have contained Vermilion Creek.

On the south slope of the range, the hogbacks and racetrack valleys that encircle Split Mountain and that flank Blue Mountain are fine examples of structurally adjusted drainage lines. Although the master streams of these areas, the Green and Yampa Rivers, were able to ignore geologic structures, many of their less competent tributaries were forced into courses adjusted to rock hardness. Many of these tributaries, in fact, are so well adjusted to the structure of the underlying bedrock that the trained eye can discern the structural form from the drainage pattern alone.

Weber, Provo, and Bear Rivers

Notable stream adjustments at the west end of the Uinta Mountains are clearly shown on the old (1900) Coalville quadrangle map and are reproduced in figure 33. At one time the Weber River, which flows to Great

Salt Lake, quite apparently drained all the western part of the range, including areas now drained by the Provo, which flows to Utah Lake. The Provo River has greatly expanded its drainage basin and, hence, its discharge, at the expense of the Weber. The Upper Provo once flowed north through Rhodes Valley into the Weber. It was diverted into the present Provo River, however, when it was captured at a point about 2 miles south of the town of Kamas.

There, the present river leaves a wide, flat-bottomed valley and enters a narrow canyon en route to Heber Valley and, ultimately, Utah Lake. Its diversion was aided, perhaps, by the clogging of Rhodes Valley with late glacial outwash. Hence, the diversion probably was a geologically recent event.

Other adjustments on the Provo have occurred farther upstream. The headwaters of the present river once flowed into Beaver Creek, a tributary of the Weber, but they were beheaded by a tributary of the ancestral Provo that eroded northeastward through Pine Valley. A broad low saddle marks the site, and many motorists today pass this point on Utah State Highway 150 without realizing that they have crossed from one drainage basin to another and that waters on opposite sides of the saddle flow to outlets 65 miles apart.

Although the Weber River has thus lost nearly half its flow to the Provo, it has gained flow from the Bear in drainage adjustments that are not really momentous, perhaps, but that are important to the drainage budgets of the streams involved. Chalk Creek, a major tributary of the Weber, arises in the northwestern Uinta Mountains (fig. 34). The headwaters of Chalk Creek flow north and northwest, enter Wyoming, and then turn abruptly south and southwest, back into Utah toward the Weber. Chalk Creek's north-flowing reach is separated from the Bear River by a low divide, but its southwest reach passes through deep canyons en route to the Weber. In all probability the headwaters of Chalk Creek formerly drained into the Bear, either directly or via Yellow Creek, but they were diverted southwestward by headward erosion of lower Chalk Creek, and the point of diversion was the "elbow" in its course at the Wyoming state line. Between that point and Pine Cliff, the direction of flow has probably been reversed. Notice (fig. 34) how the tributaries upstream from Pine Cliff are "barbed"—they flow in the opposite direction from Chalk Creek itself.

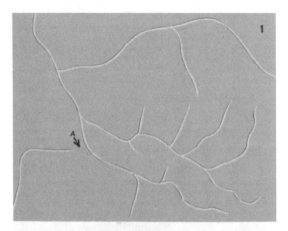

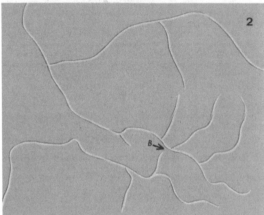

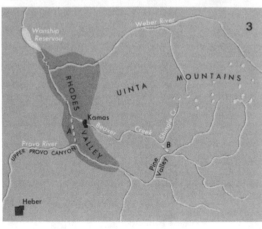

Drainage adjustments of the Weber and Provo Rivers. The headwaters of the Provo formerly drained northwest into the Weber via Beaver Creek and Rhodes Valley but were diverted west and south into the Provo by stream captures at points A and B. Heavy dashed lines indicate former drainage courses. (Fig. 33)

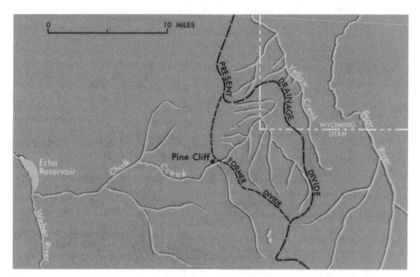

Development of Chalk Creek drainage. North-flowing tributaries of Yellow Creek, which drains into the Bear River, were captured and turned westward by Chalk Creek, which drains into the Weber River. (Fig. 34)

The Weber has also captured drainage from the West Fork of the Bear (fig. 35). A relatively minor but illustrative stream diversion north of Holiday Park characterizes a type of capture that has happened time and again elsewhere in the Uinta Mountains. Near Holiday Park, the Weber Valley is 1,200–1,500 feet lower than the West Fork of the Bear just across the divide. With much steeper gradients, the headwaters of the Weber, therefore, have a distinct hydraulic advantage over the headwaters of the West Fork. Larrabee Creek, a small steep tributary of the Weber, has eroded headward to the north, has thus intercepted the east-flowing drainage of Windy Peak, and has diverted it south from the Bear to the Weber drainage. This drainage capture was a minor physiographic event, but in a small way it typifies the ever-changing, ever-evolving sequence of events begun eons ago that culminated, from man's point of view, in the dramatic landscape of the present Uinta Mountains.

Rock Creek and the Duchesne River

Old drainage adjustments are probable between the headwaters of the Duchesne River and Rock Creek. The Duchesne River drains a well-

The Geologic Story of the Uinta Mountains

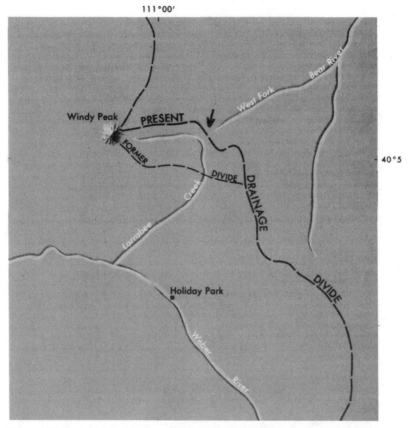

Minor stream capture at Windy Peak. Drainage down the east slope of Windy Peak, formerly to the Bear River, has been turned southward into the Weber River by Larrabee Creek. (Fig. 35)

watered area, south of the crest of the range, centered between Bald Mountain and Mount Agassiz. Drainage from this area probably flowed at one time into Rock Creek by way of the West Fork of Rock Creek (fig. 36). The evidence of capture by the Duchesne still looks convincing on topographic maps, especially on the raised relief edition of the Salt Lake City map (4 miles per inch) and on the old Hayden Peak quadrangle map (2 miles per inch), despite intensive glacial scouring since the postulated diversion. Early glaciation may even have contributed to capture by abrasively reducing the height of the intervening divide. At present, the upper Duchesne is deeply entrenched in a youthful canyon, whereas the West

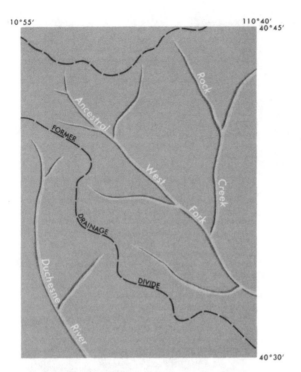

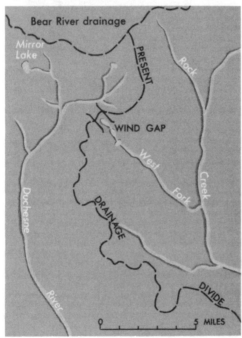

Drainage adjustments of Rock Creek and the Duchesne River. The Duchesne River has captured the headwaters of the West Fork of Rock Creek, probably in late Tertiary or early Pleistocene time. (Fig. 36)

Fork of Rock Creek heads in a broad wind gap perched above the Duchesne and open in that direction. The difference in altitude between the wind gap and the Duchesne River below is nearly 800 feet.

Glacial Diversions on the North Flank

Glaciers commonly obstruct and divert preglacial drainage lines, either by blocking them directly with ice or, on melting, by obstructing them with morainal debris. Good examples of such diversions can be seen on the north flank of the Uinta Mountains in tributaries of Beaver Creek and Burnt Fork. Diversions by glaciers of the last major advance (Pinedale) are particularly evident. The Middle Fork of Beaver Creek, for example, probably once flowed directly north into the West Fork, as shown in figure 37, but it has since been diverted eastward into its present course by massive lateral moraines left across its course by the last glacier to occupy its valley. Note also how Fellow Creek was shouldered aside by Pinedale moraines and how a tributary of Henrys Fork was diverted into Fellow Creek (fig. 37).

Hoop Lake and Lost Creek are other examples shown in figure 37. Preglacial drainage of the Hoop Lake basin was east into Burnt Fork, but it was obstructed by moraines of the former Burnt Fork glacier. The resulting depression was filled with water—thus forming Hoop Lake—and drainage from Hoop Lake spilled over into the East Fork of Beaver Creek. (Hoop Lake is a natural water body, although its level has been raised by a dam.)

Lost Creek no doubt once flowed unhindered into Burnt Fork, but its course, too, was blocked by a large moraine of the Burnt Fork glacier. Lost Creek then found an outlet in a solution cavern in the Mississippian limestone. Where the water goes from there has long been a matter of cracker-barrel speculation among old-time residents. Some guess that it emerges 15 miles to the east in Sheep Creek Canyon, where large springs issue from the Mississippian at Palisade Park. Perhaps they are right. Reportedly, roily discharge from the springs sometimes follows shower activity over the Lost Creek watershed.

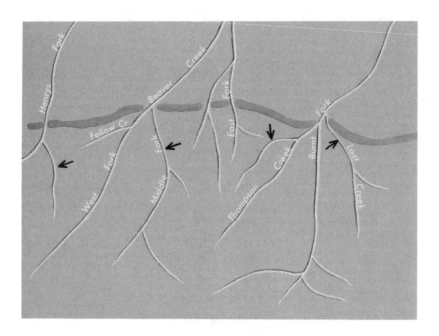

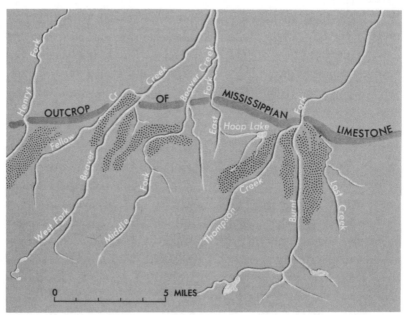

Stream diversions caused by glacial obstructions on the north flank of the Uinta Mountains. Upper diagram shows former drainage and points of diversion (arrows). Lower diagram shows present drainage and obstructing moraines. Other moraines are omitted. (Fig. 37)

WHERE BEST TO SEE	FORMATION OR GROUP	AGE IN MILLIONS OF YEARS	PERIOD	ERA
Browns Park Bishop Mountain Fort Bridger, Wyo. Uinta Basin Green River, Wyo. Flaming Gorge Reservoir Manila, Utah	Browns Park Formation and Bishop Conglomerate		Tertiary	CENOZOIC
	Duchesne River, Uinta, and Bridger Formations			
	Green River Formation			
	Wasatch Formation			
	Fort Union Formation	60		
Asphalt Ridge	Mesaverde Group	75	Cretaceous	MESOZOIC
	Hilliard, Baxter and Mancos Shales	80		
Clay Basin	Frontier Formation			
Antelope Flat	Mowry Shale	95		
Ashley Valley	Dakota Sandstone	100		
	Cedar Mtn. Formation	120		
South of Manila	Morrison Formation		Jurassic	
Dinosaur Quarry	Curtis Formation	135		
Plug Hat Butte Merkley Park Flaming Gorge	Entrada Sandstone			
	Carmel Formation	140		
Sheep Creek Split Mountain Yampa Canyon	Glen Canyon Sandstone		Triassic	
	Chinle Formation			
Horseshoe Canyon	Moenkopi Formation			
Brush Creek Canyon Wild Mountain	Dinwoody Formation			
Whirlpool Canyon	Park City Formation		Permian	PALEOZOIC
Palisades of Sheep Creek	Weber Sandstone		Pennsylvanian	
	Morgan Formation			
	Round Valley Limestone			
Gates of Lodore Red Canyon	Doughnut Shale	280	Mississippian	
	Humbug Formation			
	Deseret Limestone			
Cold Spring Mountain	Lodgepole Limestone	340		
	Lodore Formation	500	Cambrian	
Red Creek Canyon	Uinta Mountain Group		PRECAMBRIAN	
	Red Creek Quartzite	2,320		

Generalized composite section of rock formations of the Eastern Uinta Mountains. Total thickness, excluding the Red Creek Quartzite, is more than 50,000 feet. The Western Uinta section is similar, but even thicker. Age estimates are based on data by J. L. Kulp (1961), except that the Red Creek Quartzite was dated radiometrically by C. E. Hedge of the USGS. (Fig. 38)

Time and the Rocks

One can say little about the Uinta Mountains without talking about rock formations, and one cannot really understand the structure, history, or development of the range without some knowledge of the rocks. It follows that the better one understands the rocks and how they form, the better one can appreciate the grandeur and beauty of the mountains. The mountains—rising from arid basins north and south and standing starkly bare above timberline in the high interior—are clothed in dense forest only at middle altitudes. Above and below these altitudes, the mountainsides are mostly bare rock.

To many people, the term "rock formation" is ambiguous. Contrary to much popular misunderstanding, a rock formation is neither an unusual outcrop, nor a strange erosional form, nor a particular protuberance on a given piece of real estate. Rather, a rock formation is a large body of rock that has specific identifying characteristics, such as composition, color, and thickness. It may be of sedimentary, igneous, or metamorphic origin. Two or more formations of similar age or origin are sometimes lumped together as a "group," such as the Uinta Mountain Group and the Mesaverde Group.

A rock formation need not even be exposed at ground surface, but it can and generally does have lateral continuity at depth. Many formations, in fact, extend long distances under cover—that is, they pass beneath hills or basins and reappear only where brought to the surface by uplift or erosion. Some formations that are known almost entirely from drill-hole data are studied intently by specialists looking for clues to the occurrence of petroleum or other mineral resources. Thus, a limestone bed that forms a line of cliffs high in the Uinta Mountains may be penetrated by a drill many miles away and thousands of feet underground in one of the adjoining basins. A geologist might visit the mountains to learn about formations that are deeply buried many miles away. Many rock formations that crop out in the Uinta Mountains are recognized throughout much of the western interior of the United States.

Rock formations of the Uinta Mountains range widely in age, composition, and appearance. The oldest are more than 2 billion years old; the youngest are just now being detected in lakes and streams. Figure 38 shows diagrammatically the sequence of rock formations in the middle and eastern parts of the range. Many particulars are omitted from this chart, and the reader should turn to more technical publications for the details. Moreover, rocks vary appreciably from place to place. Some formations in one part of the range are lacking in other parts. Most of the formations of the Uinta Mountains are sedimentary and were deposited on flat beds as broad blankets of sand, gravel, clay, or limy mud. Solidified, such deposits became sandstone, quartzite, conglomerate, shale, and limestone, with many gradations between. Though they may be only a few hundred or a few thousand feet thick, these formations may extend laterally for hundreds or thousands of square miles.

Travelers crossing the range on Utah State Highway 44 between Vernal and Manila have an excellent opportunity to acquaint themselves with the formations that make up most of the drainage. From the flanks inward, progressively older rocks are exposed, somewhat like the rings of a tree. Recognizing the unique value of this section, the USDA Forest Service and the Utah Field House of Natural History have erected markers along the road, identifying each formation in turn.

The same formations, but without labels, are widely exposed elsewhere in the range, at such places as Dinosaur National Monument, Vermilion Creek, and Flaming Gorge, and with much variation along the Weber and Duchesne Rivers. With some training, anyone can quickly identify the major rock units exposed throughout the range. In particular, such units as the Uinta Mountain Group, the Deseret Limestone, the Weber Sandstone, the Glen Canyon Sandstone, the Morrison Formation, and the Dakota Sandstone lend special character and distinction to the Uinta Mountain scene. These units, however, are but half a dozen of more than 30 recognized formations in the Uinta Mountains.

Rocks of the Uinta Mountains are of many ages, ranging from Precambrian to Quaternary. Two distinctive Precambrian formations are present: the older Red Creek Quartzite and the younger Uinta Mountain Group. The Red Creek is confined to a few square miles in the northeastern part of the range, but the Uinta Mountain Group is exposed over many hundreds of square miles, from one end of the range to the other

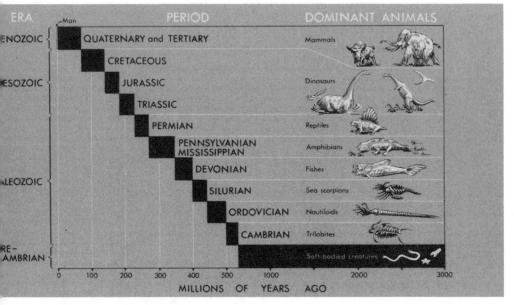

ERA		PERIOD	DOMINANT ANIMALS

Man

ENOZOIC — QUATERNARY and TERTIARY — Mammals

MESOZOIC — CRETACEOUS / JURASSIC / TRIASSIC — Dinosaurs

PERMIAN — Reptiles

PENNSYLVANIAN MISSISSIPPIAN — Amphibians

PALEOZOIC — DEVONIAN — Fishes

SILURIAN — Sea scorpions

ORDOVICIAN — Nautiloids

CAMBRIAN — Trilobites

PRE-AMBRIAN — Soft-bodied creatures

0   100   200   300   400   500   1000   2000   3000

MILLIONS OF YEARS AGO

Major divisions of geologic time

and across the crest from flank to flank. All the high peaks consist of this rock. In brief, its name is well chosen.

Cambrian rocks crop out along the south flank of the range, but they were eroded from the north and northeast flanks before Mississippian time. Ordovician, Silurian, and Devonian rocks are lacking, although rocks doubtfully regarded as Devonian are exposed in a few places. Each subsequent geologic period is represented by one or more rock formations.

The Old Uinta Trough

Throughout most of recorded geologic time, long before the mountains themselves appeared, the Uinta Mountains region was occupied by a slowly subsiding basin—a broad elongate trough flooded much of the time by shallow-marine waters. This trough was an arm, or embayment, of a great seaway that intermittently covered western North America from Arctic Canada to Mexico. At times the seaway connected the Arctic Ocean to the Gulf of Mexico. Gentle flexing of the earth's crust caused repeated incursions and withdrawals of marine waters. Weight from accumulating sediments that were washed in from the sides of the trough helped cause the sea floor to subside.

The Uinta trough generally subsided more toward the west than toward the east. Marine conditions were therefore more prevalent toward the west, and deposits accumulated more rapidly. As a result, most formations thicken toward the west but thin to the east, north, and south. The present Uinta Mountains, uplifted by mountain-forming movements in the crust 70± million years ago, are relative latecomers to the scene.

Precambrian Time:
4.6 Billion–570 Million Years Ago

The earliest, most complex sequence of events recorded in the Uinta Mountains took place more than 2 billion years ago, at a time so remote that it boggles the mind. The details of what happened, even though they are incompletely known, are too involved to be reiterated here, but they are described elsewhere (Hansen, 1965, p. 22). In summary, the Red Creek Quartzite was deposited on a well-washed shore. Made up mostly of clean sand, but including some mud and limy mud, it was then compacted, uplifted, folded, metamorphosed, and faulted. Sandstone, shale, and limestone were transformed by heat and pressure deep within the earth into quartzite, mica schist, and marble. Molten igneous rock was then injected into fractures and bedding planes, where it cooled and solidified and, finally, was transformed by renewed heat and pressure into hornblende schist, or amphibolite (fig. 39). New minerals were recrystallized from old, including such showy varieties as actinolite, garnet, kyanite, and staurolite.

Thus was formed the Red Creek mountain range, long since disappeared. The mountains trended east to west, much like the present Uintas, but, no doubt, they were larger and higher. The age of the mountains, 2.3 billion years, has been determined by radiometric dating—that is, by measuring the decay products of radioactive minerals whose decay rates are known physical constants. Erosion then reduced the Red Creek mountains to their roots—a nearly flat plain on which the Uinta Mountain Group began to accumulate. The base of the Uinta Mountain Group

Contorted Red Creek Quartzite, more than 2 billion years old, intruded by dark igneous dike. Exposure is about 100 feet high. (Fig. 39)

and the roots of the old mountains are well exposed north of Browns Park in a great unconformity visible from the floor of the valley.

When the Uinta Mountain Group began to accumulate, possibly a billion years ago, mountains of Red Creek Quartzite still stood to the northeast not far away. Bouldery alluvial fans spread out into the Uinta trough and formed coarse conglomerates that now are preserved in the Uinta Mountain Group north of Browns Park. The trough subsided slowly, and its rate of subsidence was counterbalanced almost exactly by the accumulation of sediment. No doubt the weight of the sediment was a factor in causing the subsidence. Ripple marks, fossil mud cracks, and even rain prints suggest shallow tidal pools and desiccating mudflats. True marine conditions existed in the western part of the range, where C. A. Wallace has identified an ancient shoreline. At the top of the Uinta Mountain Group, in the western part of the range, a thick shale formation called the Red Pine may have accumulated well offshore. Altogether about 24,000 feet of sandstone, shale, and conglomerate accumulated in the old Uinta trough before gentle uplift ended Precambrian deposition and ushered in the Paleozoic Era.

Paleozoic Time: 570–245 Million Years Ago

The Uinta Mountain area once again was an expanse of hills, this time composed of red sandstone, quartzite, and shale. It was exposed to erosion until Middle to Late Cambrian time, when the sea moved in again. The sea encroached from west to east, depositing first the Tintic Quartzite and Ophir Shale at the west end of the range and then, as it spread eastward, depositing the Lodore Formation toward the east. The Lodore, well exposed in Dinosaur National Monument, is a succession of red, gray, and green sandstone and shale containing brachiopods and rare

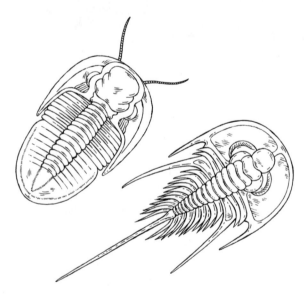

Trilobites,
about natural size.

trilobites, primitive little cousins of our modern horseshoe crab. These fossils are the oldest clear evidence of life in the Uinta Mountains.

Cambrian time was followed again by emergence of the land. Just when this happened is uncertain because there is a stratigraphic gap of about 150 million years; Ordovician, Silurian, and probably Devonian rocks are lacking. At any rate, marine Mississippian rocks (Lodgepole Limestone) overlap the eroded edges of the Cambrian in many parts of the range. Along most of the present north flank, Cambrian rocks were completely eroded away before Mississippian time.

The Mississippian Period brought in a reexpansion of the seaway and, with it, widespread deposition of the Madison Limestone and its equivalents. These rocks crop out from Alberta to Arizona, even to Sonora. The famous Redwall of Grand Canyon is the southern correlative. In the Uintas these rocks comprise four formations: the Lodgepole, Deseret, Humbug, and Doughnut. The Mississippian limestones form bold outcrops wherever they are exposed. They form the imposing "gateways" in Blacks Fork, Henrys Fork, Beaver Creek, and Burnt Fork, the "Palisades" of Sheep Creek (fig. 40), the east wall of Irish Canyon, and the giant stair-steps of Whirlpool and Lodore Canyons. Though resistant to erosion, these limestones are highly susceptible to solution by carbonatic waters, and many caverns have formed along their outcrops. Large springs issue from some of them, as at Sheep Creek. At other places whole streams disappear into the ground. Lost Creek, on the north

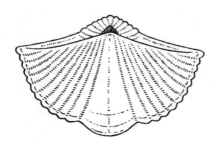

Spirifer, a common late Paleozoic brachiopod, about natural size.

Palisades of Sheep Creek. Eroded edge of upturned Mississippian limestone in contact with the Uinta Mountain Group, left. (Fig. 40)

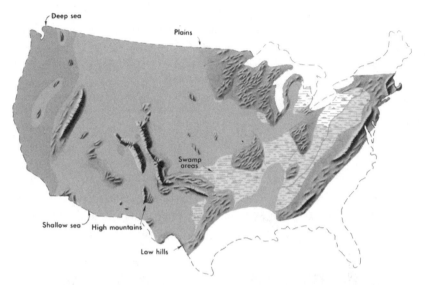

Generalized geographic map of the United States in Middle Pennsylvanian time. (Fig. 41)

flank near Burnt Fork, is such a stream; it empties into a sinkhole at the base of the Lodgepole Limestone. Dry Fork of Ashley Creek, Pole Creek (a tributary of the Uinta), Brush Creek, and Little Brush Creek all lose drainage into the Mississippian limestone.

In addition to limestone, rocks of Mississippian age in the Uinta Mountains include sandstone and shale. Marine organisms flourished, and their fossilized remains are abundant, especially corals, brachiopods, crinoids, and snails.

Throughout the remainder of Paleozoic time, the restless sea ebbed and flowed. The formations succeeding the Mississippian rocks—the Round Valley, Morgan, Weber, and Park City—all contain marine fossils. Yet, deposition was interrupted from time to time as the sea floor emerged. High mountainous islands stood to the southeast in Colorado and far to the west in Nevada in late Paleozoic time (fig. 41). The Round Valley Limestone was deposited well offshore, and the Morgan Formation was deposited near shore in shallow water of variable depth. The Weber Sandstone is mainly a beach deposit, and some of it may have accumulated as onshore sand dunes. But it also contains fossiliferous marine limestone. Much more could be said about the Weber; it is one of the most impressive and picturesque formations in the entire Uinta Mountains.

The Park City Formation, on top of the Weber, is varied most of all. It contains a host of rock types laid down under diverse conditions. Deep water lay west of the Uintas; a shoreline lay east. Limestone predominates in most places and makes long resistant hill slopes (fig. 53), but phosphate rock is the most distinctive. It formed by the upwelling of organic-rich marine waters on the margin of a long-vanished continental shelf. These deposits are now being mined in big open pits near Vernal for use in the production of phosphatic fertilizer. The Park City Formation also contains gypsum, sandstone, conglomerate, and mudstone, cherty nodules and geodes, and, locally, many fossils.

Mesozoic Time: 245–66.4 Million Years Ago

The Paleozoic Era drew to a close with clear marine waters still spread across the Uinta Mountains region. Following in Triassic time, environmental conditions continued to change. At first, shallow marine muds of various shades of gray, red, and green were spread across the region, perhaps in a broad tidal delta, shoaling toward the east (the Woodside, Dinwoody, and Moenkopi Formations). As time passed, the sea floor subsided toward the west, and marine limestones (Thaynes Limestone) again began to accumulate. A few clams and their cousins lived in the Western Uinta region at that time, and countless ammonites teemed in the waters farther west. But in the Eastern Uinta region, gypsum was forming from concentrated sulfate brines in shallow arms of the sea, and few organisms survived. These gypsum beds are in the lower and middle parts of the Moenkopi Formation, laterally equivalent in the eastern part of the range to the Thaynes Limestone in the west. The gypsum is interbedded with ripple-marked red mudstone; together they form a conspicuous pinkish band several hundred feet thick in the Moenkopi north of Vernal and south of Manila (fig. 42).

Regional uplift followed deposition of the Moenkopi Formation. Thousands of square miles of eastern and southern Utah, northern Arizona, western Colorado, and southern Wyoming were gently elevated,

Triassic rocks exposed in Sheep Creek Canyon near Flaming Gorge. Gray Dinwoody Formation (foreground) overlain by Moenkopi, Chinle, and Glen Canyon, forming canyon wall and skyline. (Fig. 42)

draining off the sea. No mountains were formed, but much of the Moenkopi was carried away by erosion (fig. 43). The succeeding Chinle and Ankareh Formations (lateral equivalents, in part) began to form first as gravelly river-channel deposits, filling hollows and swales in the eroded top of the Moenkopi, then as a complex assemblage of red, purple, ocher, and green silts, muds, sands, and gravels, augmented by occasional falls of ash from distant volcanoes. Conditions clearly were nonmarine; fossil wood and remains of reptiles, amphibians, and mollusks, though uncommon, all suggest a terrestrial, perhaps marshy environment. Utter desiccation then ensued, and the whole western interior of the country became a vast desert.

The Chinle and Ankareh Formations extend upward into the Glen Canyon* Sandstone, a thick deposit of pink and gray to orange fossil sand dunes, which advanced across the old Chinle-Ankareh landscape.

* The name "Glen Canyon" has been applied in the Uinta Mountains to rock formations called Nugget or Navajo, a usage that obviates certain correlation problems with other regions.

Moenkopi Formation (lower right) beveled by erosion and overlain unconformably by Chinle Formation (upper left). One mile southwest of Flaming Gorge. (Fig. 43)

The Glen Canyon is one of the prominent cliff-forming units in the Uinta Mountains, and it enhances the local scenery wherever it crops out. Northwest of Vernal it forms the walls of Dry Fork Canyon (fig. 44) and the "beehives," arches, and alcoves of Neal Dome. Near Manila the Glen Canyon forms the heights of Jessen Butte, the north wall of Sheep Creek Canyon, the caprocks of Flaming Gorge, and the spine of Dutch John Mountain. Equivalent rocks are widespread in the canyon lands of southern Utah, where, for example, they form the towering walls of Zion and Glen Canyons. Altogether, an immense expanse of dunes covered much of the western interior of the country. Though centered in Utah, the dunes extended into Nevada, Arizona, New Mexico, Colorado, Wyoming, and Idaho. The crossbedded internal structure of the Glen Canyon Sandstone, a clear indication of its origin, is well preserved in the solid rock and gives the formation much of its scenic character.

Deposition of the Glen Canyon Sandstone ended when the interior seaway readvanced across the region, again from the west. The sea bottom deepened to the west, like the earlier pattern of encroachment, and it shoaled to the east. Desert conditions persisted along the shore as the water lapped eastward. Accordingly, thick gray limestones interbedded with gray and red shales west of the Uinta Mountains diminish in thickness and pass gradually eastward into a much thinner section of red, green, and gray mudstone, shale, gypsum, and subordinate limestone. The thick section is called the Twin Creek Limestone. Its thin eastward counterpart is called the Carmel Formation. The Carmel disappears at about Skull Creek near the east end of the Uinta Mountains.

The Twin Creek sea was well populated with shellfish, and their remains are easy to find. Some limestone beds contain abundant fossilized clams, oysters, ammonites, sea urchins, and crinoids. Near Flaming Gorge, shell beds, or "coquinas," consist almost entirely of tightly packed shells and shell fragments. North of Vernal, beautifully preserved fossil clam shells have been replaced—through the process of petrifaction—by reddish-orange jasper. Conditions for the survival of life were unfavorable, however, when or where gypsum and red mud were accumulating in the Carmel Formation; these deposits, of course, contain no fossils.

After the Twin Creek and the Carmel were deposited, the sea withdrew again. Its withdrawal this time was recorded by the Entrada Sandstone. The Entrada consists of clean white to pink cliff-forming sandstone interbedded with red or brown mudstone. The red or brown beds, locally called the Preuss Sandstone, may have accumulated in shallow stagnant water, but the white to pink sandstones were dunes. As the sea withdrew the dunes crowded close against the shore, and as the sea readvanced the dunes pulled back. Mudstone thus predominates toward the west along the mountains, and sandstone, toward the east. In the more easterly sections, the Entrada consists entirely of windblown sand. East of the pinchout of the underlying Carmel Formation, the Entrada and the Glen Canyon combine to form a single thick body of cliff-forming sandstone.

The withdrawal was short-lived, for the sea returned again, readvancing far beyond its previous shoreline to central Colorado. Its deposits this time were gray to greenish-gray marine sandstone, shale, and lime-

Cliff-forming Glen Canyon Sandstone, in Dry Fork Canyon near Vernal, Utah. Photograph by Clint McKnight. (Fig. 44)

stone, called the Curtis Formation. (In the Weber River drainage, it is generally called the Stump Sandstone.) The Curtis sea had a diverse fauna. Among the more typical inhabitants was the belemnite, a soft-bodied relative of the squid and cuttlefish, having an internal shell of "cuttlebone," commonly preserved as a fossil. Several kinds of clams, snails, and, locally, ammonites were also abundant. Near Manila, count-less little brachiopods left their shells in the uppermost beds of the formation.

Squidlike belemnite, about half natural size.

As the Curtis sea pulled back, nonmarine mud, sand, and gravel from distant sources to the west or south built up a thick alluvial plain, now the Morrison Formation. Frequent falls of volcanic ash must have dimmed the hot Jurassic sun. The Morrison Formation is famed espe-cially for its dinosaurs, nowhere better exemplified than in Dinosaur National Monument. But it also is one of the West's leading uranium producers, though not in the Uinta Mountains. Chief production from the Morrison is in southwestern Colorado and southeastern Utah.

The Morrison Formation is very widespread, and its variegated red, gray, green, and lavender clays are everywhere characteristic. It crops out around mountainous uplifts throughout most of the Rocky Mountain region and passes deep underground in between. Dinosaur bones from the Uinta Mountains have been known for many years. Powell noted "reptilian remains" during his explorations, and O. A. Peterson of the American Museum of Natural History identified dinosaur bones in 1893. The world-famous deposit north of Jensen was discovered in 1909 by Earl Douglass of the Carnegie Museum. Carnegie Museum operated the quarry from that time to 1922. The United States National Museum, the American Museum of Natural History, and the University of Utah oper-ated it in 1923–24. After that time, collecting was discontinued. Assem-bled specimens are on display in museums in Denver, Lincoln, New York,

Pittsburgh, Salt Lake City, Toronto, and Washington. In all, 26 nearly complete skeletons and parts of about 300, representing 10 different species, were recovered before the quarry was shut down (Good and others, 1958). Meat eaters as well as plant-eating dinosaurs, large and small, were taken from the quarry. Among the more noteworthy were *Diplodocus,* about 76 feet long; *Apatosaurus,* about 70 feet long; and *Stegosaurus,* about 20 feet long—all vegetarians—and *Allosaurus,* a bipedal carnivore, about 30 feet long.

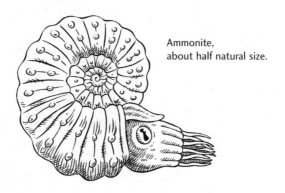

Ammonite, about half natural size.

Small dinosaurs were collected also, some only the size of chickens. A nearly complete lizardlike adult crocodile found by the Carnegie Museum was only about 7 inches long. In addition, the quarry has yielded the remains of large crocodiles and of turtles, freshwater clams, and plants—surely one of the most extraordinary fossil deposits in the world.

The miracle of fossilization is thus reaffirmed by the dinosaur quarry. The odds against such a deposit being preserved are almost astronomical. The likelihood that the remains of a particular creature will be preserved, especially a land-dwelling animal, is extremely remote. At best, only the hard parts, such as the bones, are apt to be preserved. When, for example, one contemplates the millions of bison that roamed the Great Plains less than 200 years ago and the present scarcity of their remains, one can appreciate the odds against fossilization. Of the trillions of organisms that have lived and died on the earth, only the smallest fraction has been left behind as fossils.

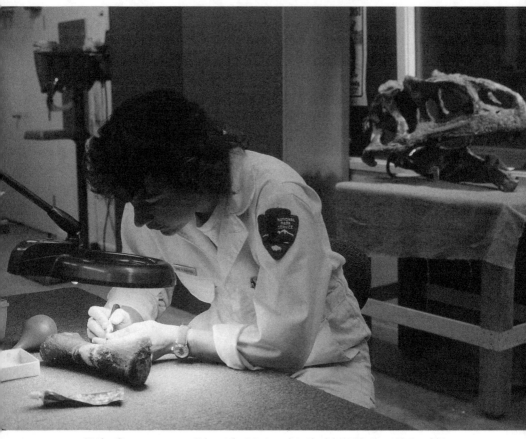

At the dinosaur quarry. Paleontologist at work in the lab at Dinosaur National Monument. Note *Allosaurus* skull in background. Photograph by C. Kane. (Fig. 45)

Quick burial is essential, to remove the remains from the weather. Bones left to the sun and elements soon disintegrate; wood left lying in the forest soon returns to the soil. Petrifaction then must replace or reinforce the organic tissues with stone—generally silica—molecule by molecule to forestall destruction during subsequent earth movements or rapid deterioration on reexposure to the elements. Finally, reexposure must be at the right place and at the right time for man to discover the fossils before they are destroyed by weathering and erosion.

Today, the National Park Service has developed the dinosaur quarry as an in-place exhibit. Bones that were carefully uncovered from their stony matrix have been left in place, so that the visitor can see them just as they were deposited. Research has shown that the dinosaurs did not all die together in some great catastrophe. Their remains accumulated one by one in the bed of an ancient river, as their dead bodies were washed downstream from time to time. Sediment rapidly covered the deposit, protecting the bones from decay. Lithification followed, encasing the bones in solid rock. Mountain-forming uplift and erosion finally exposed the deposit as Douglass found it 150 million years later.

As the Morrison episode was drawing to a close, events far to the west of the Uinta Mountains were shaping the depositional pattern of the future. High mountains were rising in California, and were soon to rise in Nevada and western Utah, shedding their sediments eastward. The seas, which through much of Paleozoic and Mesozoic time had invaded the region from the west, were pushed to the east; the newly formed lands to the west were never again flooded. Nonmarine Lower Cretaceous rocks accumulated on the Morrison Formation without any clear-cut interruption in deposition. These rocks are called the Cedar Mountain Formation. They have a varied content and variegated color, like the Morrison, but they are characterized especially by soft mudstones in pale-lavender tints. In many places they contain "gastroliths," the so-called gizzard stones of dinosaurs. (Whether or not these were ever actually ingested by dinosaurs is open to question. Authorities do not agree.) The base of the Cedar Mountain is indicated in some places by a pebbly ledge-forming unit called the Buckhorn Conglomerate Member. The Buckhorn is well exposed, for example, just south of Escalante Overlook, along the road from Dinosaur National Monument Headquarters to Harpers Corner, but, where the Buckhorn is lacking, even experts are hard put to find the precise top of the Morrison or the bottom of the Cedar Mountain.

On top of the Cedar Mountain is the Dakota Sandstone. This formation is also very widespread. It extends from Montana to New Mexico and from central Utah to the Great Plains. The Dakota was deposited on the shore of a reexpanding seaway, which encroached this time from the east as part of an inland sea connected to the Gulf of Mexico. As the sea again spread across the shore, sands and gravels accumulated on the beach. The Dakota thus contains marine as well as nonmarine sandstone, conglomerate, and coal-bearing shale that must have accumulated in

Fossil ripple marks. Together with mud cracks, these features indicate deposition in intermittent shallow water. (Fig. 46)

coastal swamps. Here and there are bits of petrified wood. Many fossil ripple marks and mud cracks indicate intermittent wetting and drying (fig. 46).

Because it is resistant to erosion, the Dakota Sandstone forms bold outcrops on the flanks of the mountains. Where its dip is gentle, it forms sloping bedrock plains (cuestas) bounded by cliffs facing toward the mountains. Where its dip is steep, it forms narrow rocky ridges (hogbacks) that may extend for miles without a break.

Subsidence of the crust and a flood of ocean water brought Dakota time to a close and introduced the overlying Mowry Shale. This distinctive dark-gray sea-bottom deposit weathers silvery gray. Most of it was derived from airborne volcanic ash that settled into the water. The ash had a high silica content that gave the shale a hard porcelainlike quality, one of its distinctive attributes. Inasmuch as the shale thickens greatly to the northwest of the Uinta Mountains, its volcanic source may have been in that direction. Another attribute of the Mowry Shale—a really diagnostic one, and an unmistakable clue to the identity of the formation—is the presence of countless well-preserved fish scales found with very little effort on nearly every outcrop. Powell called the formation the "teleost shale" in allusion to the ill-fated donors of the scales. Perhaps the fish were killed by ash falls. The Mowry sea contained, as well as fish, a few oysters and, at times, some very large ammonites. The largest of these were more than 2 feet across.

At the close of Mowry time, the sea pulled back again, and the bottom temporarily emerged. Then, with a few pulsations back and forth, the sea returned decisively for a long time. During the pulsating phase, the Frontier Formation was deposited. This formation, like the Dakota, consists of beach-laid sandstone, near-shore marine sandstones, and coastal-swamp deposits. Throughout most of the Uinta Mountains the Frontier Formation is twofold: a slope-forming shale unit below, and a cliff-forming sandstone above. Oysters and other oysterlike shellfish are very characteristic, but the faunal list includes many other mollusks, also, particularly in the shaly lower part of the formation. A coal zone near the top of the shale yielded a few tons of coal in pioneer days, when nearness of supply was more important than the tonnage available. (In other parts of the West, such as Kemmerer, Wyoming, and Coalville, Utah, the Frontier contains large reserves of coal.) The upper sandstone unit of the Frontier, where it is not faulted out or concealed by overlapping rocks, forms a hogback on both flanks of the Uinta Mountains. Travelers on U.S. Highway 40 drive beside it for many miles along the south side of Blue Mountain; at Dinosaur National Monument it crops out just behind Monument Headquarters. The road across Blue Mountain, from Monument Headquarters into Harpers Corner and Echo Park, incidentally, provides excellent views of the whole geologic section of the Eastern Uinta Mountains, from the Frontier Formation down to the Uinta Mountain Group.

As the Frontier Formation accumulated in the Uinta Mountains area, high, newly formed mountains stood not far to the west, perhaps just west of the present Wasatch Range. Inasmuch as these mountains were the source of the sediments, the Frontier Formation thickens enormously in that direction. In the Eastern Uinta Mountains, it is only 100–200 feet thick, but at Coalville, just west of the Uintas, the Frontier is more than 4,400 feet thick and contains coarse conglomerate, as well

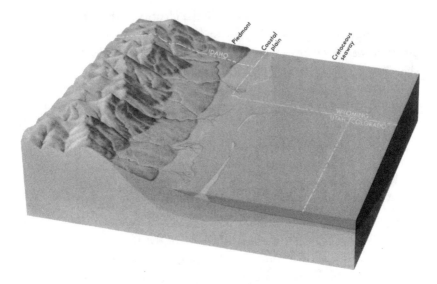

Northeastern Utah and adjacent areas in Late Cretaceous time. High, newly formed mountains in western Utah shed debris eastward. Coastal-plain deposits accumulated between the mountains and the old Cretaceous seaway, and the crust subsided in the piedmont belt. (Fig. 47)

as sandstone, shale, and coal. A few miles farther west, near Salt Lake City, it is more than 8,700 feet thick. The Late Cretaceous geography of northeastern Utah can thus be visualized as follows: A broad shallow sea extended eastward from the foot of high north-trending coastal mountains that stood some distance west of the present Uintas (fig. 47). Gravelly alluvial fans reached east from the mountains to the shore. Waves and shore currents constructed sandy beaches and offshore bars. Dense junglelike forests grew down to lagoons behind barrier beaches. The waters teemed with shellfish.

The buildup of high mountains to the west perhaps depressed the crust in front of them, for the crust is not rigid, and it responded to the added weight. The sea deepened, and the Frontier Formation thinned eastward, giving way to deep-water clays and muds that later became the Mancos Shale and its equivalents.

The Mancos Shale and its equivalents are very thick deposits of soft, dark-gray shale. They weather into shades of light gray or tan and form dreary badlands on both flanks of the eastern part of the range. On the north flank the names Hilliard and, locally, Baxter Shale are used.

Though the tops and bottoms are not exactly equivalent, all three units—Mancos, Hilliard, and Baxter—were deposited in the same sea and were once continuous bodies of the same clay.

Where not eroded, the Hilliard, Baxter, and Mancos have a thickness of several thousand feet, a maximum of more than a mile. This is not to say that the water was ever a mile deep. The sea bottom subsided slowly as the mud accumulated, and the crust was depressed beneath it. The shale thins to the west, intertonguing in that direction with sandstones and conglomerates of the thickening Frontier Formation.

Geologic events at this time became very complex throughout western Utah but remained fairly simple in the area that was to become the Uinta Mountains. The Mancos seaway gradually withdrew to the southeast—not, however, without many short-lived readvances (fig. 48). Long tongues of beach sand interlayered with coal-bearing coastal swamp deposits marked its retreat and introduced the Mesaverde Group, which overlies and intertongues with the Mancos (fig. 49). Sandstone, shale, and coal comprise the Mesaverde Group, and geologists recognize many distinctive subdivisions. Different names, therefore, have been applied to the Mesaverde in various parts of the West, but for our purposes, the name Mesaverde Group will suffice.

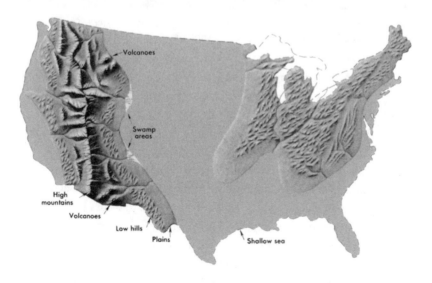

Generalized geographic map of the United States in Late Cretaceous time. (Fig. 48)

The Glades. Hogbacks formed from resistant sandstone of the Mesaverde Group, northeast of Flaming Gorge. Antelope Flat (left) is underlain by soft Hilliard Shale. (Fig. 49)

By Mesaverde time, the old seaway that had dominated the geography of the Uinta Mountains area since Precambrian time was yielding, finally and irrevocably, to mountain-building forces stirring again to the west. The Mesaverde is the youngest group of rocks that predates uplift of the Uinta Mountains. If anything younger was deposited, it was stripped away by erosion as the mountains began to rise.

The whole Rocky Mountain System was taking shape at about this time. Mountain ranges were forming in response to forces in the earth's crust that were directed horizontally, obliquely, and vertically. Some ranges began to rise before others. At the same time, large basins were forming between the mountain ranges, where the crust buckled down instead of up. Many of these began to fill with water and sediment. Some parts of the crust were stressed to the point of rupture, and they failed by fracturing, or faulting, as great blocks of rock moved past one another, particularly at the flanks of the mountains. Compression dominated in some places, and when it finally slackened, uplifted masses began to subside, in response to gravity. By Cenozoic time, most of the Rocky Mountain ranges and basins as we know them today—not all, but most of them—were well delineated.

Cenozoic Time: 65 Million Years Ago to the Present

The Tertiary Period and the Beginnings of the Modern-Day Landscape

The rise of the Uinta Mountains in Late Cretaceous time was guided by the east-trending old Uinta trough and the thick prism of sediments that had accumulated through the ages. Near the west end of the range, the Evanston Formation, of Late Cretaceous and Paleocene age, contains conglomerate that seems to have been shed from the Uinta Mountains as erosion attacked the newly formed heights. This conglomerate thins toward the west away from the Uinta Mountains, and coarsens toward the east (T. E. Mullens, oral communication, 1968). It contains quartzitic cobbles, moreover, that probably came from the Uintas. Similar conglomerate accumulated in the Currant Creek area near the Duchesne River.

Deposits of latest Cretaceous age are lacking in the Eastern Uinta Mountains and may never have been deposited, but strong uplift and deep erosion plainly had taken place by earliest Tertiary (Paleocene) time. Large-scale faulting had thrust the mountainous core of the range thousands of feet up over the flanks, especially near Flaming Gorge. The Fort Union Formation near Flaming Gorge was deposited across the eroded edges of the older pre-Tertiary formations. Besides carrying the impressions of Paleocene leaves (oak, hickory, sycamore, and others) the Fort Union contains pebbles derived from the new Uinta Mountains to the south, including shale chips that contain fish scales from the Mowry Shale. By this time erosion must have penetrated at least 8,500 feet into the cover of the Uinta arch. By Wasatch time (Eocene) the Eastern Uinta Mountains were breached to their quartzitic core, and a flood of quartzite pebbles, cobbles, and boulders eroded from the Uinta Mountain Group was deposited in alluvial fans at the foot of the mountains. These fans became the Wasatch Formation, a widespread accumulation of sandstone, conglomerate, shale, and limestone in variegated shades of red and gray. North of Manila, the Wasatch contains enormous boulders of Paleozoic limestone—some 8 feet or more across—that were carried off the mountains by Tertiary flash floods or mudflows.

Meanwhile, as the mountains rose and faulting continued, the adjacent basins subsided. Large freshwater lakes formed to the north and south, filling the basins and lapping against the mountains. The complex history of these lakes—a story in itself—cannot be told here. It has been studied intensively by many geologists, most particularly by W. H. Bradley (1948, 1964). In aggregate, the lake deposits are called the Green River Formation, famous for its vast reserves of oil shale, its enormously large deposits of sodium salts (trona beds), and its remarkably well-preserved fossils. The accumulations of limy muds on the lake bottoms proved to be an ideal medium for the preservation of fossils. Fish remains from the Green River Formation have become world famous, not only for their excellent preservation (fig. 50), but also for their abundance. They

A well-preserved Eocene fish, *Diplomystus dentatus,* collected from the Green River Formation. About one-third natural size. Photograph by W. T. Lee. (Fig. 50)

The Geologic Story of the Uinta Mountains

are displayed in museums throughout the world. In addition, wonderful impressions of leaves, flowers, fruits, insects, a feather, and even the remains of a bat have all been collected from the Green River Formation.

Most lakes are relatively short-lived geologic features. They are filled to overflowing with sediment; they are drained dry by erosion at their outlets; or their basins are destroyed by crustal warping. The Green River lakes were no exception. They varied greatly in extent and depth from time to time. Saline phases occurred when lake levels dropped below their outlets. At high stages, overflow around the east end of the mountains, according to Bradley, may have joined the Green River Basin to the Uinta. Inevitably, as deltal and alluvial deposits encroached upon the basins, the lakes finally disappeared. They were succeeded by the Bridger and Uinta Formations, north and south. These formations—soft mudstone and sandstone mixed with volcanic ash (fig. 51)—attracted the attention of early-day collectors for their rich faunas of Eocene mammals

The Bridger Formation, typically exposed in badlands southeast of Fort Bridger, Wyoming. (Fig. 51)

and reptiles, including the bizarre many-horned *Uintatherium*. Another formation in the Western Uinta Basin, the Duchesne River Formation, also contains remains of early Tertiary reptiles and primitive mammals.

By Oligocene time the Uinta Mountain scene was beginning to have a fairly modern appearance. The present divide probably had taken shape. Broad sloping plains (pediments) were forming on the flanks. An apron of coarse gravel, the Bishop Conglomerate (of uncertain age but either Oligocene or Miocene), was accumulating at the outer margin of the plains. Thick remnants of this conglomerate still cap high mesas along the edge of the mountains. Large-scale faulting, propelled by gravity, was beginning in the eastern part of the range, dropping the eastern crest relative to the flanks. The Browns Park Formation was beginning to accumulate; this, perhaps, in Miocene time.

The stage was now set for the sequence of events that shaped the present scene—the development of modern drainage across the mountains, the cutting of the great river canyons, and the glacial sculpture of the high peaks—events already discussed in the section on landscapes. All these events helped mold the Uinta Mountains as we know them today. All past geologic events, in fact, helped shape the present landscape, even the deposition of the Red Creek Quartzite more than 2 billion years ago. Clearly, the landscape, the rocks, and the processes that formed them both are not entities unto themselves. They are interdependent parts of an integrated whole, a continuum of matter, space, and time.

Uintatherium was about 5 feet high at the shoulder.

The Geologic Story of the Uinta Mountains

Geologic Structure

Structural geology deals with the relations of rock masses to one another, particularly insofar as the rocks have been deformed by forces within the earth. Studying the geologic structure helps the geologist unravel the history of the earth.

Most of the early-day geologists who visited the Uinta Mountains commented on the unusual structural relation of the range to its neighbors. Trending due east 150 miles from the little town of Kamas, the range is directly athwart the northerly trend of the neighboring ranges and uplifts of the Rocky Mountains. The crestline of the Uinta Mountains actually is broadly arcuate, trending about N. 10° E. from Kamas to about the center of the range near the Burro Peaks, then swinging gradually east-southeastward to its eastern terminus near the Little Snake River. Structurally and topographically, the Uinta Mountains intersect the north-trending Wasatch Range on the west at an abrupt 90° angle.

On the east, however, the trend of the Uinta Mountains veers to the southeast and merges with the northwest-trending folds of the White River Plateau and the Park Range. All the major structural trends in that part of the Rocky Mountains and the adjacent Colorado Plateau, for that matter, are northwest, including most major faults. So, the only really major discordance of the Uinta Mountains is along the west margin of

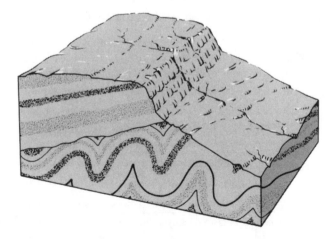

An unconformity. Ancient folded rocks eroded and overlain by tilted younger rocks.

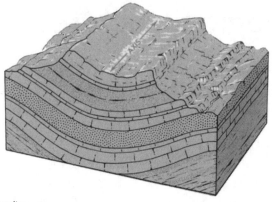

An eroded syncline.

the Rocky Mountains, particularly at the Wasatch juncture.

In trying to visualize the crustal movements that produced the Rocky Mountains 60–70 million years ago, one is bemused by the magnitude of the uplifting forces that raised thousands of cubic miles of rock in the Uinta Mountains alone from a position near or even below sea level to present altitudes, thousands of feet higher. If the strata that have been eroded from the Uinta Mountains were restored in our imagination, the upper layers would be 30,000–40,000 feet above sea level. This is not to say that the mountains ever stood that high, even though some early-day geologists surmised that they did. For as the mountains rose, erosion attacked and carried away the heights, and the higher the mountains rose, the more vigorous the attack. It is doubtful that the mountains were ever really much higher than now—possibly a few thousand feet higher, but never Himalayan, certainly, in size.

The upbuckling that produced the mountains was accompanied by comparable downbuckling under the basins. As the mountains rose, the basins subsided, so that deposits once near sea level throughout the region are now 12,000–13,000 feet high in the mountains but are as much as 30,000 feet below sea level beneath the Green River and Uinta Basins. Moreover, debris shed from the mountains accumulated in the adjacent basin, and its added weight further depressed the basin floors. Obviously, the crust is not unyielding; it responds to pressures directed horizontally, vertically, or obliquely. Such pressures—building up at the end of the Cretaceous Period—caused the crustal movements that produced the Uinta Mountains.

Folds

In the High Uintas, the crestline is much closer to the north flank than to the south. The marked asymmetry of the range is related to the structure of the underlying bedrock. In brief, the Uinta Mountains are carved from an immense anticlinal fold whose outline closely coincides with the outer limits of the mountains. The uplifted strata of the range, therefore, slope outward from the crest of the fold toward the flanks, so that successively younger rocks are crossed as one travels outward from the interior of the range toward the flanks. The Uinta anticline is compound in that its crest has two main apexes or culminations—one in the Western Uinta Mountains and one in the Eastern Uintas—something like two peanuts in one shell. In a general way, however, the crest of the fold is nearly flat, so that the bedding or stratification along the crest is almost horizontal, from Mount Watson on the Provo-Weber divide all the way to Douglas Mountain near the east end of the range. Away from the fold axis the dip of the strata steepens, first gradually, then abruptly toward the flanks (figs. 52 and 53).

Flat-lying strata in the Uinta Mountain Group (Precambrian) at crest of Uinta Mountains. Wilson Peak, altitude 13,053 feet. (Fig. 52)

Dipping strata on north flank of Uinta Mountains at Sheep Creek Canyon. Dipslopes and cliffs in foreground are eroded from the Park City Formation. (Fig. 53)

Like the mountains themselves, the Uinta anticline is asymmetrical. Its crest is much closer to the north flank than to the south, therefore the strata on the north flank dip more steeply than the strata on the south. Severe bending along the flanks, moreover, has led to rupture and large-scale faulting, particularly along the north flank.

In the west half of the range, the crest of the anticline very nearly coincides with the crest of the range—in some places it is slightly north, in others, slightly south. Thus, despite the removal of many thousands of feet of rock by erosion, the position of the main divide apparently has remained almost unchanged since the fold came into being 70-odd million years ago.

The Uinta anticline is complex. Elongate east and west domes share a common axis, or crestline, which curves gently to the north in a broad arc. The east and west domes are almost mirror images of each other—comparable in size and shape—although the east dome is appreciably more complex than the west. Each dome is 70–75 miles long and 30–35 miles across. Both are faulted internally, and both are partly bounded by faults.

Associated with the east dome, but not the west, are several large subsidiary folds, known since the time of Powell; they include the Blue Mountain, Split Mountain, arid Section Ridge anticlines and their synclinal counterparts. These are shown in figure 54. In addition, there are many smaller folds, a few miles long, on the flanks of the larger ones. Most of these have been actively prospected for petroleum, and among them are several economically important oil fields, the most significant recent discovery being that of the Bridger Lake field, just west of Henrys Fork, high on the north flank. Bedrock at Bridger Lake is concealed by thick glacial deposits, and the oil field was discovered by purely geophysical means. Its discovery in 1965 set off a flurry of exploration and leasing activity that is still in progress.

Geologic cross sections through the Eastern Uinta Mountains. p€, Precambrian rocks; €, Cambrian; M, Mississippian; P℗, Permian and Pennsylvanian; J℞, Jurassic and Triassic; K, Cretaceous; T, Tertiary. Note that Cambrian rocks are missing from the north flank. Vertical scale x 2. (Fig. 54)

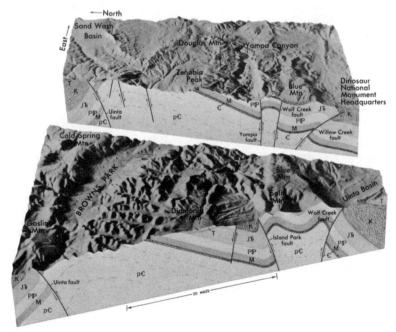

Faults

Mention has been made of the asymmetry of the Uinta anticline. The crest of the fold is much nearer the north flank than the south; consequently, the strata slope, or dip, more steeply down the north limb of the fold than down the south (fig. 54). To a large extent, bending on the north limb proceeded to the point of rupture. Even solid rock will yield to long-sustained stress. As the folding progressed, the strata were first tilted, then overturned, and, finally, broken. Deep crustal movements pushed the Uinta anticline northward over the margin of the Green River Basin, causing the development of three large faults: the North Flank fault to the west, the Henrys Fork fault at the center, and the Uinta fault to the east. These are reverse, or "thrust," faults; the fault surfaces dip steeply under the mountains, and the upthrust, mountainward block overrides the basinward block. Of the three faults, the Uinta appears to be the largest, having a displacement of about 34,000 feet, or the distance one block moved past the other. The North Flank fault may have a displacement of as much as 25,000 feet; the Henrys Fork fault, perhaps more than 12,000 feet.

Large faults also broke the south limb of the anticline, although none of these appears to have displaced the strata as much as those on the north limb. From west to east, they include the South Flank fault, extending 80 miles east from the Duchesne River area to Ashley Creek, the Deep Creek fault zone northwest of Vernal, and the Willow Creek, Island Park, Mitten Park, Wolf Creek, and Yampa faults of Dinosaur National Monument. These faults have generally been regarded as "normal faults," faults in which the surface of rupture slopes under the downdropped block. Recent studies by several investigators, however, suggest that some of them are "reverse" faults, at least at depth. In other words, the mountainward, upthrown side moved up and over the basinward, downthrown side. In the cross sections (fig. 54) the Yampa, Island Park, Wolf Creek, and Willow Creek faults are shown as reverse faults; the South Flank fault is shown as a normal fault.

Most of the large faults of the Uinta Mountains are fairly broad zones of disturbance rather than simple fractures. There is much ground-up and shattered rock, much distortion, and many subsidiary fractures. The Uinta fault, largest in the range, serves to illustrate.

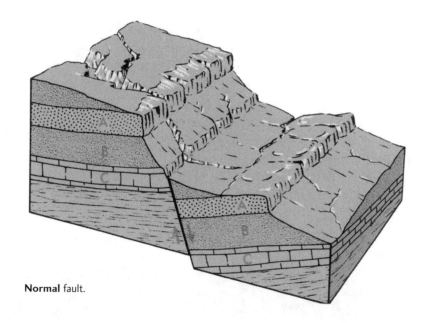

Normal fault.

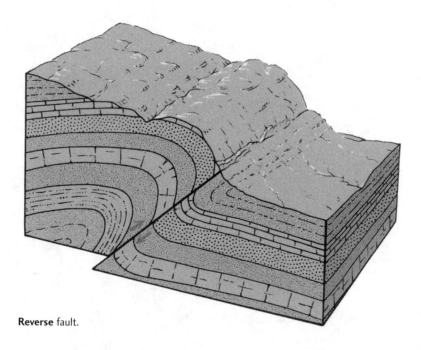

Reverse fault.

The Uinta fault was named and first described by Powell (1876, p. 177). A more detailed description can be found in U.S. Geological Survey Professional Paper 490 (Hansen, 1965). The fault is well exposed and readily accessible; the best places to see it are just north of the town of Dutch John and just south of Clay Basin. Figure 55 shows the fault at Hideout Canyon, before Flaming Gorge Dam was built. From its terminus southwest of Manila, the Uinta fault trends sinuously east-northeast for about 45 miles as a continuous fracture. It increases in displacement and width to about the 109th Meridian, where it passes beneath the Browns Park Formation. About 10 miles to the east, it emerges as the

Uinta fault from rim of Hideout Canyon. Weber Sandstone (left and center) is upturned against Uinta Mountain Group. (Fig. 55)

Sparks fault and is virtually continuous for another 20 miles. Discontinuous fractures extend at least another 30 miles.

The Uinta fault has many subsidiary branches. It also contains large slices of rock torn from the walls and dragged into the fault zone. One such slice north of Dutch John is more than 2 miles long—possibly 3 miles—and as much as 400 feet across. Well back from the fault surface the strata have been tilted up on edge and even overturned, but where Tertiary rocks are involved, they are little deformed despite the severe deformation in the immediately underlying Cretaceous rocks. Apparently, the fault had more than one period of movement: the first and strongest movement in Late Cretaceous or earliest Tertiary time severely dislocated the involved rocks; a second movement, after the Tertiary rocks had been deposited across the fault, simply sheared off the Tertiary strata with little other effect. Still later, gravitative movements accompanied the collapse of the Eastern Uinta Mountains (pgs. 60 and 102).

Some further mention must be made of the faults of Dinosaur National Monument. These faults are especially significant for their dramatic physical expression and for the part they had, therefore, in shaping the impressive local scene. Few faults anywhere are better displayed. All of them happen to be eroded to levels in the earth's crust where contrasting strata help disclose their geometric form. Thus, between Round Top Mountain and Echo Park in the heart of the monument, the Weber Sandstone is repeated topographically in four huge escarpments, or stairsteps, bounded from rim to river by the Yampa, Red Rock, and Mitten Park faults and the canyon of the Yampa River itself. The fault relations are unmistakably clear.

The faults of the Dinosaur National Monument area further disclose sharp flexing of the strata in their walls, again dramatically exposed (fig. 56). Like an expression of frozen movement, or of time standing still, these faults accent the grandeur of the scene and stir wonder in the heart of the viewer. Even the most languid boater, lazing in the hot summer sun, is duly awed as he floats past the Mitten Park and Island Park faults, upstream and down, respectively, from Whirlpool Canyon.

Internally, the Uinta Mountains are also extensively faulted, and few large expanses of rock can be seen that do not have some faulting. Two major zones are especially noteworthy. One zone, named the Crest fault by J. D. Forrester (1937, p. 645), extends west to east for many miles

Sharply dragged strata of the dramatic Mitten Park fault along the Green River below Harpers Corner. (Fig. 56)

along the west dome of the anticline just north of the crest (fig. 57). Recent mapping by Crittenden, Wallace, and Sheridan (1967, plate 1) showed that the Crest fault is a complex zone of many interrelated fractures with various trends and offsets. Neither the total displacement nor the direction of movement is known with certainty, although the north side of the zone, according to Crittenden, Wallace, and Sheridan probably has dropped about 2,000 feet.

The other zone extends along the east dome of the anticline, also north of the crest, from the vicinity of Red Canyon to Browns Park, and to the east end of Cold Spring Mountain. This fault zone was instrumental in the collapse of the east end of the Uinta Mountains in Tertiary time. Augmented by gravitative movements on the Uinta fault to the north, and in concert with movement on the Yampa fault to the south, it led to the collapse of the Uinta Mountain "graben," noted on page 60. In summary, the crestal part of the east dome collapsed under gravity relative to the flanks of the dome some time after compressive mountain building had ended. When the compressional forces that arched up the dome relaxed, gravity began to pull the dome back down. Total collapse along the crestline amounted to about 4,500 feet and, thus, accounts for most of the difference in altitude between the east and west halves of the range (Hansen, 1965, p.172).

Crest fault, a zone of fractures, viewed from the head of Blacks Fork of the Green River. Note flat strata (left of fault zone) and dipping strata (right). Photograph by Max D. Crittenden, Jr. (Fig. 57)

Goslin Mountain. A tilted remnant of the Gilbert Peak erosion surface on the north flank of the range. Now sloping southward (left), it formerly sloped northward. (Fig. 58)

Crustal Warping

Gravitative movements of the earth's crust began in the Uinta Mountains before the Browns Park Formation was deposited, even before the Gilbert Peak erosion surface was cut. These movements continued until after deposition of the Browns Park Formation was completed. Along the north side of Browns Park the strata were fractured, tilted, and dragged into near-vertical attitudes. Locally, they were even overturned, as in Jesse Ewing Canyon, near the west end of Browns Park. Elsewhere, collapse was accompanied by large-scale regional tilting of the ground surface, clearly shown by deformation of the Gilbert Peak erosion surface on the north flank of the range (fig. 58), by tilting of correlative surfaces on the south flank, and by tilting of strata, such as the Browns Park Formation. Partly bounded by faults, different areas were tilted in different directions. But the tilting on the north flank was predominantly inward toward the south, and on the south flank, predominantly inward toward the north, just as one would expect of a foundering mountain range.

Quaternary Crustal Movements

How long warping and faulting continued in the Uinta Mountains after the Browns Park Formation was deposited is unknown, although both processes clearly extended into the Quaternary Period and may still be in progress. Little direct evidence of Quaternary deformation has been observed, but careful search perhaps would disclose more. Some evidence is available in Lodore Canyon. After the Green River had entrenched itself 1,500 feet or so into Lodore Canyon, it apparently stopped cutting downward and began to widen its valley. But the river was then rejuvenated, and it resumed cutting downward an additional 600–800 feet to its present depth. Thus was formed the two-tiered valley, shown in profile (p. 46). Although the cause of rejuvenation is unclear, renewed uplift is suspect, and it might have happened in Quaternary time. Canyon profiles suggest similar rejuvenation upstream in Red Canyon, but the evidence is not compelling.

In the Uinta Basin at Towanta Flat, just south of the mountain front and just west of Lake Fork, local but clear-cut deformation occurred in late Pleistocene time. A graben and several subordinate fault scarps more than 3 miles long displace what probably are Bull Lake age (late Pleistocene) glacial-outwash terraces, but the faults do not cut across younger terraces or moraines of Pinedale age. One of the fault scarps is about 40 feet high.

On the north flank of the range just north of Phil Pico Mountain a probable faulted alluvial fan was noticed by Max Crittenden, Jr. (oral communication, 1968), on aerial photographs. The age of this feature is uncertain, but it probably is no younger than the Bull Lake glaciation.

On the Middle Fork of Blacks Fork just south of the North Flank fault are the remains of a spectacular rockfall avalanche. Thousands of tons of rock, now spread over an area of perhaps 100 acres, broke free from the tilted outcrop of the Mississippian limestone and plummeted 1,000 feet to the canyon floor. There, part of the rock avalanche crossed the Middle Fork and traveled 200 feet up the opposite canyon wall. Just when this event happened is unknown, but the evidence is very fresh, and it must have happened in the not-too-distant past—a few hundred years ago, at most. Although the avalanche might have been released by gravity alone, it is the sort of feature that commonly is associated with earthquakes, such as the 1959 Madison River slide in Montana. Because of nearness of the avalanche to the North Flank fault, its release could easily be ascribed to geologically recent activity on the North Flank fault.

Historically, however, the Uinta Mountains have been seismically quiet, even though many earthquakes are recorded annually in the Wasatch Range to the west. But the historical period has been too short geologically to permit valid projections of events in the Uinta Mountains, either into the past or into the future.

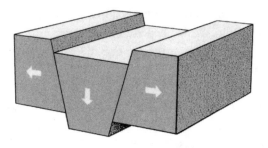

A simple graben formed by tension.

What the Future Holds

We can affirm, however, that changes will indeed continue. Slowly, perhaps, but inexorably. The deep canyons, the mountain peaks, and the "everlasting" hills—though firm they stand—are only bit players acting out their roles on the endless stage of time. John Wesley Powell would agree.

Each summer, hikers will hear the reports of loosened rocks, crashing shotlike from the ridge lines into the valleys far below. Each winter, unseen avalanches will sweep more rocks down from the heights. Each spring, freshets will scour the beds of creeks and rivers. And each fall, when new ice coats the ponds and fresh snow powders the summits, fingers of frost will again reach deep into the rocks, searching out each crack and crevice.

Each year, landslides and mudflows will carry more debris to the bottom lands, leaving new scars on the hillsides. In time these will heal and leave no trace. Through the years, lakes and ponds will slowly gather silt. First, pond lilies, then rushes, grasses, herbs, and, finally, willows will encroach upon the shores, converting some lakes to ponds, some ponds to marshes, and some marshes to meadowlands. Some meadowlands, gullied and dissected by erosion, will become badlands. Some of these, perhaps, will grow into canyons.

Drainage adjustments will follow the patterns of the past; some streams will capture the headwaters of others. Aided by the hand of man, some changes will outpace all those gone before. And with wisdom, man's way will be the right way. If man's way prevails, snowmelt from the High Uintas will moisten the parched fields of central Utah. Powell would approve.

Much of the high country—unchanged wilderness since the time of Powell and now protected by law—will remain inviolate for the enjoyment of generations yet unborn. Powell would applaud.

Glossary

amphibolite: a rock-forming, silicate mineral.

antecedence: the act of going before or preceding.

anticline: a fold in the rock strata inclining downward on both sides from a median line.

conglomerate: sedimentary rock consisting of pebbles held in a fine-grain matrix.

cross bedding: patterns in sandstone formed by shifting winds or waters during deposition.

fault: a place of fracture and relative movement of rock layers.

felsenmeers: any of several types of deposits of loose, rocky rubble accumulated through weathering over time. Include talus, protalus, and rock glaciers.

formation: an identifiable rock type.

group: a division of stratified rocks consisting of two or more formations.

gypsum: hydrated calcium sulfate; a soft mineral used in construction of wallboard.

hornblend schist: dark, iron-rich metamorphic rock created through heat and pressure from fine-grained sedimentary rock.

igneous: rock formed from solidified magma.

lacustrine: formed at the bottom or along the shore of a lake.

limestone: sedimentary rock composed of marine deposits.

lithology: physical characteristics of a rock or stratigraphic unit.

marble: limestone changed and hardened through metamorphosis.

metamorphic: rock formed from previously existing rocks through heat and/or pressure.

mica schist: a crystalline, metamorphic rock that easily separates into thin, transparent laminae.

monocline: strata dipping in only one direction.

moraine: linear deposit of rock fragments left by a glacier.

mudstone: clayey rock with the texture and composition of shale.

mya: million years ago.

protalus: ridgelike deposit of loose, rocky rubble accumulated at the foot of a snowbank or ice field. Also commonly and accurately called a *terminal moraine.*

quartzite: metamorphic rock consisting of quartz in interlocking grains.

reverse fault: one that leaves an overthrust above the downthrown block.

sandstone: sedimentary rock formed from sand deposits.

schist: a crystalline, metamorphic rock with a relatively parallel grain pattern.

sedimentary: rock formed by compression and cementation of sediment.

shale: sedimentary rock comprised of clay.

stream capture: event whereby the course of a stream changes to follow a newly formed drainage.

superposition: existing over or above another position.

syncline: a fold in the rock strata inclining upward on both sides from a median line.

talus: sheetlike or conelike deposit of loose, rocky rubble accumulated at the base of a steep slope or cliff.

unconformity: a major break in a sequence of rock strata that represents a period when no new sediments were laid down or where earlier sediments were eroded away.

References

The following short list of references refers mostly to authors cited directly in the text. The geologic literature on the Uinta Mountains is very extensive, and no attempt has been made here at completeness. A few papers of general interest or wide scope not directly cited in the text, however, have been included. All of these papers are technical or semi-technical in content. Most of them can be seen in any big city library. Some of the government publications are available from the superintendent of documents, although most of them are out of print.

Atwood, W. W., 1909, Glaciation of the Uinta and Wasatch Mountains: U.S. Geological Survey Professional Paper 61, 96 pages.

Bradley, W. H., 1936, Geomorphology of the north flank of the Uinta Mountains [Utah]: U.S. Geological Survey Professional Paper 185-1, pages 163–199.

Bradley, W. H., 1948, Limnology and the Eocene lakes of the Rocky Mountain region: Geological Society of America Bulletin, volume 59, number 7, pages 635–648.

Bradley, W. H., 1964, Geology of the Green River Formation and associated Eocene rocks in southwestern Wyoming and adjacent parts of Colorado and Utah: U.S. Geological Survey Professional Paper 496-A, 86 pages [1965].

Capps, S. R., 1910, Rock glaciers in Alaska: Journal of Geology, volume 18, pages 359–375.

Crittenden, M. D., Jr., Wallace, C. A., and Sheridan, M. J., 1967, Mineral resources of the High Uintas Primitive Area, Utah: U.S. Geological Survey Bulletin 1230-1, 27 pages.

Emmons, S. F., 1877, Descriptive geology: U.S. Geological Exploration 40th Parallel (King), volume 2, 890 pages.

Forrester, J. D., 1937, Structure of the Uinta Mountains: Geological Society of America Bulletin, volume 48, number 5, pages 631–666.

Good, J. M., White, T. E., and Stucker, G. F., 1958, The Dinosaur Quarry, Dinosaur National Monument, Colorado-Utah: Washington, D.C., [U.S.] National Park Service, 47 pages.

Hansen, W. R., 1965, Geology of the Flaming Gorge area, Utah–Colo-Wyo: U.S. Geological Survey Professional Paper 490, 196 pages.

Hayden, F. V., 1871, Preliminary report [fourth annual] of the United States Geological Survey of Wyoming and portions of contiguous Territories (being a second annual report of progress): U.S. Geological Survey Territories, 4th Annual Report, 511 pages.

Kinney, D. M., 1955, Geology of the Uinta River–Brush Creek area, Duchesne and Uintah Counties, Utah: U.S. Geological Survey Bulletin 1007, 185 pages.

Kulp, J. L., 1961, Geologic time scale: Science, volume 133, number 3459, pages 1105–1114.

Marsh, O. C., 1871, On the geology of the eastern Uinta Mountains: American Journal of Science, 3d series, volume 1, pages 191–198.

Merrill, G. P., 1906, Contributions to the history of American geology: Washington, D.C., U.S. National Museum, Annual Report of the Smithsonian Institution, 1904, pages 189–733.

Powell, J. W., 1876, Report on the geology of the eastern portion of the Uinta Mountains and a region of country adjacent thereto: U.S. Geological and Geographical Survey Territories (Powell), 218 pages.

Richmond, G. M., 1965, Glaciation of the Rocky Mountains in Wright, H. E., Jr., and Frey, D. G., eds., The Quaternary of the United States: Princeton, New Jersey, Princeton University Press, pages 217–230.

Ritzma, H. R., 1959, Geologic atlas of Utah, Dagget County: Salt Lake City, Utah, Utah Geological and Mineralogical Survey Bulletin 66, 111 pages.

Sears, J. D., 1924, Relations of the Browns Park formation and the Bishop conglomerate and their role in the origin of the Green and Yampa Rivers: Geological Society of America Bulletin, volume 35, pages 279–304.

Sears, J. D., 1962, Yampa Canyon in the Uinta Mountains, Colorado: U.S. Geological Survey Professional Paper 374-1, 33 pages.

Stokes, W. I., and Madsen, J. H., Jr., compilers, 1961, Geologic map of Utah—northeast quarter: Salt Lake City, Utah, Utah Geological and Mineralogical Survey, scale 1:250,000.

Untermann, G. E., and Untermann, B. R., 1954, Geology of Dinosaur National Monument and vicinity, Utah-Colorado: Salt Lake City, Utah, Utah Geological and Mineralogical Survey Bulletin 42, 228 pages.

Van Hise, C. R., 1892, Correlation Papers; Archean and Algonkian: U.S. Geological Survey Bulletin 86, 549 pages.

Wahrhaftig, Clyde, and Cox, Allan, 1959, Rock glaciers in the Alaska Range: Geological Society of America Bulletin, volume 70, pages 383–436.

Walton, P. T., 1944, Geology of the Cretaceous of the Uinta Basin, Utah: Geological Society of America Bulletin, volume 55, number 1, pages 91–130.

White, C. A., 1889, On the geology and physiography of a portion of northwestern Colorado and adjacent parts of Utah and Wyoming: U.S. Geological Survey 9th Annual Report, pages 677–712.

Woolley, R. R., 1930, The Green River and its utilization: U.S. Geological Survey Water-Supply Paper 618, 456 pages.

Index

(Italic page numbers indicate major references)

G

M

Y

Z

About the Author

Now retired from the U.S. Geological Survey, Wallace Hansen is considered to be the unequivocal expert on the geology of the Uinta Mountain region. As a schoolboy, he made his first trip to the Uintas in his uncle's 1928 Essex coupe. Later, he made many trips of his own in his 1932 Chevrolet roadster, hiking to the many lakes to fish for trout. In 1940 he visited the Uinta Mountains as a student. Under the direction of R. E. Marsell and the late Hyrum Schneider, the University of Utah set up a geologic field camp on the Duchesne River. Working from there, he gained his first real appreciation of the country's impressive geology.

While in the employ of the USGS in the 1950s, Mr. Hansen returned to the Uintas to map a block of quadrangles along the Green River between Flaming Gorge and Browns Park. Over his many years, Mr. Hansen became intimately familiar with the Uinta Mountains and its impressive geologic history.

Among Mr. Hansen's many published reports, maps, and books are *The Black Canyon of the Gunnison, in Depth* (Southwest Parks and Monuments Association, Tucson, AZ), *Effects of the Earthquake of March 27, 1964 at Anchorage, Alaska* (USGS and National Academy of Science, Washington, DC), and *Greenland's Icy Fury* (Texas A&M University Press, College Station, TX). Hansen holds the U.S. Department of the Interior's highest honor, the Distinguished Service Award (1979), the Rocky Mountain Association of Geologists' Outstanding Scientist Award (1991), and the American Association of Petroleum Geologists' Journalism Award (1995).